CORE ORGANIC CHEMISTRY

CORE ORGANIC CHEMISTRY

Joshua Howarth
Dublin City University

JOHN WILEY & SONS
Chichester · New York · Weinheim · Brisbane · Singapore · Toronto

National 01243 779777

International (+44) 1243 779777
e-mail (for orders and customer service enquiries):
cs-books@wiley.co.uk.
Visit our Home Page on http://www.wiley.co.uk.
or http.//www.wiley.com

Other Wiley Editorial Offices

John Wiley & Sons, Inc., 605 Third Avenue,
New York, NY 10158-0012, USA

WILEY-VCH Verlag GmbH, Pappelallee 3,
D-69469 Weinheim, Germany

Jacaranda Wiley Ltd, 33 Park Road, Milton,
Queensland 4064, Australia

John Wiley & Sons (Asia) Pte Ltd, 2 Clementi Loop £02-01,
Jin Xing Distripark, Singapore 129809

John Wiley & Sons (Canada) Ltd, 22 Worcester Road,
Rexdale, Ontario M9W 1L1, Canada

British Library Cataloguing in Publication Data

A catalogue record for this book is available from the British Library

ISBN 0 471 98333 0

Typeset in 10/12pt Times by Vision Typesetting
Printed and bound in Great Britain by Biddles, Guildford and King's Lynn
This book is printed on acid-free paper responsible manufactured from sustainable forestation, for which at least two trees are planted for each one used for paper production.

CONTENTS

PREFACE

In order to obtain the highest qualification you can at university, in order that you may carry on along the career path you wish, you have to know how to do five things well; these are:

You must know **what is required** in your course.
You must **understand** the subject matter you have been lectured on.
You must grasp the **fundamentals**.
You must **learn** the subject matter you have been lectured on.
You must **practice**.

This book attempts to help you achieve all five of these requirements with respect to organic chemistry. The book is not a classical textbook, it is primarily a detailed syllabus and revision aid; it sets out the **requirements** for **fundamental** organic chemistry. The book covers 95% of the subject matter taught first in the vast majority of universities. The **understanding** of any subject is inevitably dependent on the person studying; however, access to study material that has simplicity of explanation and directions to further information is invaluable; this book has tried to incorporate these two requirements. Once an understanding of the subject is achieved, **learning** it requires good notes and discipline. This book provides a comprehensive set of notes akin to those you would take from your lecture course; it is a **revision aid**–the discipline you must find yourself. Part of learning is **practice**, and often the hardest part of practice is finding enough questions of the correct standard to practise on. The appendix of this book contains **one hundred and seventy** questions from university exams to practise on.

So who is the book aimed at?

The new student studying for a chemistry degree.
The student taking organic chemistry as part of another degree.
Any student wishing to revise their fundamentals.

This book **does not** cover any type of spectroscopy course (UV, NMR, MS or IR). There is an excellent summary of these techniques in Field *et al.* (1995) and in Williams and Flemming (1995). Also this book does not cover nomenclature of organic compounds; this topic is covered in the classical organic chemistry texts.

Many thanks to those who have helped me complete this book, especially the proof readers: Paul Glynn, Padraic Kenny, Kevin Gillespie, Jason Finnegan, Michaela Walshe and Frances Weldon. Thanks to Chris Richards for encouragement whilst I was writing the book. Also many thanks to those universities which have allowed me to use their exam questions in the appendix of this book.

Josh Howarth

1 GENERAL ORGANIC CHEMISTRY

1.1 Comments on Representation of Molecules

There are a number of ways of representing the molecules we study in organic chemistry. At the simplest level, representations of organic compounds show all the atoms in the molecule and their bonds: for example,

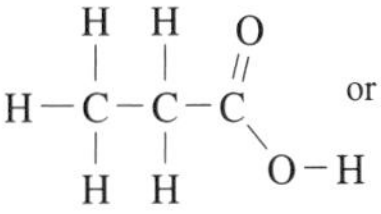

$$H-\underset{H}{\overset{H}{C}}-\underset{H}{\overset{H}{C}}-C(=O)-O-H \quad \text{or} \quad CH_3\text{-}CH_2\text{-}CO_2H \qquad CH_3\text{-}CH_2\text{-}CH_2-\overset{CH_3}{\overset{|}{C}}H\text{-}CH_3$$

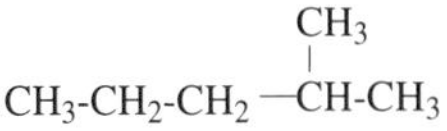

Frequently, such representations are abbreviated by:

1. Omission of many carbon–hydrogen and carbon–carbon bonds, as shown below.
2. Condensation of group formulae, as also shown by the examples below.

$$CH_3CH_2 = C_2H_5 \qquad COOH = CO_2H$$

It is worth noting that abbreviation in this way is limited to **small** groups. With larger groups ambiguity rapidly arises, for example C_3H_7 could be *normal*-propyl $CH_3CH_2CH_2$ or *iso*-propyl CH_3CHCH_3. Note that *sec*, *n*, *i* (or *iso*) and *t* (or *tert*) are used to denote *secondary*, *normal*, *iso* and *tertiary* groups.

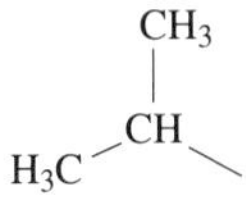

$$H_3C-\overset{CH_3}{\overset{|}{CH}}- \qquad \textit{iso}\text{-propyl}$$

3. Consolidation of groups of the same kind; examples are given below.

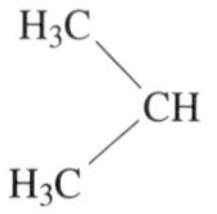 = $(CH_3)_2CH$ $CH_3CH_2CH_2CH_2$ = $CH_3(CH_2)_3$

Further abbreviation can take the form of replacing groups by an abbreviation of their names.

Me = CH_3 (methyl) Et = C_2H_5 (ethyl) Ph = C_6H_5 (phenyl)

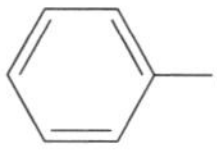 $—CO_2H$ = $PhCO_2H$ (benzoic acid)

The ultimate abbreviation is found in the "stick" formulae, in which all carbon atoms and their attached hydrogen atoms are omitted. The representation is constructed of bonds joining carbon atoms, together with hetero atoms (O, N, P, S, *etc.*) and the bonds joining these to the basic skeleton.

$CH_3CH_2CH_2CH(CH_3)CH_3$ = 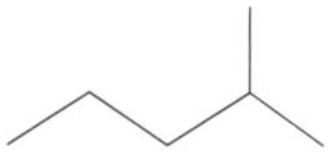

$CH_3CH(CH_3)CH_2CH_2CH(CH_2CH_3)CH_2CHO$ = 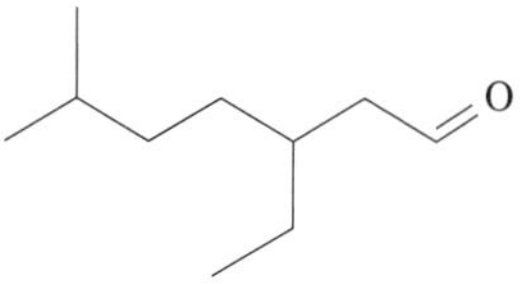

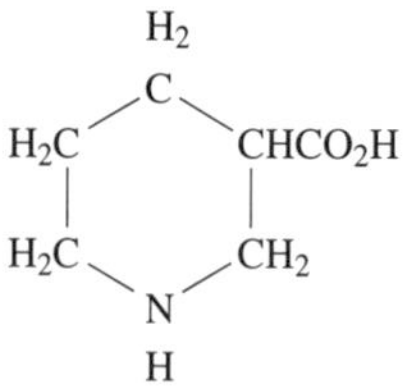

=
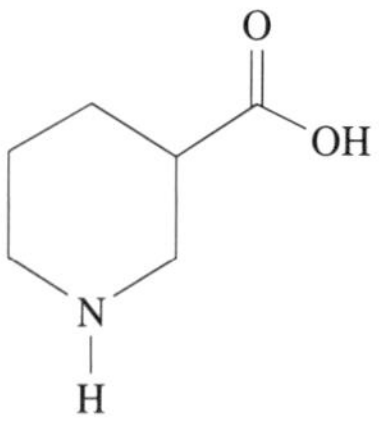

$H_3C-CH{=}CH-CH_2-C{\equiv}C-CO_2H$ = 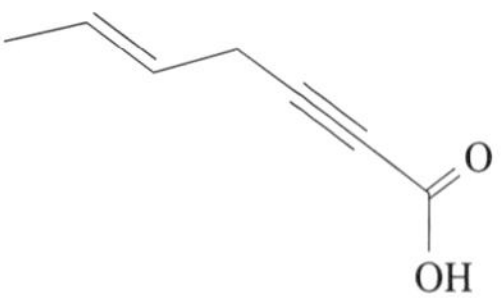

Note that the aldehydic hydrogen is not shown in the second example above. This hydrogen is sometimes drawn in. Also in the last example above notice that a triple bond is drawn in a linear fashion.

This is the most economical and widespread method of representing structures. The enormous advantage gained by an early grasp of this method of representation cannot be underestimated.

Many representations of molecules use a wedge (or a bold line) to show bonds coming out of the plane of the paper and a hashed wedge (or a dotted line or hashed bond) to indicate bonds lying behind the plane of the paper. When the bond can be either behind or in front of the plane of the paper a wiggly line or normal bond representation is used. The dotted line is also used to indicate partial bonds in transition states, see 3.41 for an example. The representations shown here are used further on in this book.

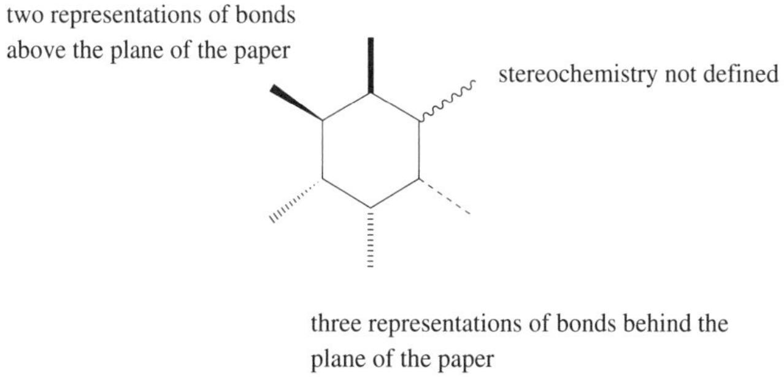

two representations of bonds above the plane of the paper

stereochemistry not defined

three representations of bonds behind the plane of the paper

Chemists often interchange the words "proton" and "hydrogen" when describing the group "H" in a molecule; both are acceptable.

1.2 A Note on Fischer Projections

(See Gunstone 1974, although out of print it is worth acquiring or reading.)

There is a method of representing molecules as vertical and horizontal lines. These representations are known as Fischer projections. They are most often encountered in carbohydrate chemistry, but have been largely superseded by three-dimensional drawings.

	A	B	C	D
top	CHO	CHO	CHO	CH_2OH
left–right	H — OH	H — OH	HO — H	H — OH
bottom	CH_2OH	CH_2OH	CH_2OH	CHO

If a Fischer projection is to be correctly used certain rules must be obeyed:

1. The structure is written in a vertical rather than a horizontal form and the carbon atom with the lower number (this is in nomenclature terms; in the example shown above this is CHO rather than CH_2OH) is written uppermost.
2. In Fischer projections the groups occupying the vertical positions are thought of as lying behind the plane of the paper. Those occupying horizontal positions are projecting above the plane of the paper. So A could, for example, be represented by B.
3. For the purposes of comparison a Fischer projection may be rotated through 180° in the plane of the paper. The projection must be kept within the plane of the paper. Thus C can be represented by D, but this is not the same as A.

Fischer projections are considered again with respect to chirality, see 1.24 and 1.26.

1.3 List of Functional Groups

In order to converse in the language of chemistry it is of **primary importance** to know the twenty or so common functional groups. No progress will be made unless these can be identified in a molecule. The following list provides a quick reference for some of these functional groups, R = an alkyl or aryl group.

R—X alkyl or aryl halide; R—OH alcohol; R—SH thiol; $R—NR_2$ amine; R—O—R ether

R—C≡C—R alkyne; R—C≡N nitrile; R—N≡C isonitrile; $R_2C{=}CR_2$ alkene

aldehyde; ketone; carboxylic acid; ester; amide

acyl halide; anhydride; ketal; acetal; epoxide

1.4 Molecular and Structural Formulae

A compound is generally considered to be known if the following two are known:

1. The **molecular formula**, which is the number of each atom in a given compound, for example C_3H_8O.
2. The **structural formula**, which is the sequence in which the atoms are bonded together in the compound.

It is possible to work out the empirical molecular formula if the percentages of each element present in the compound are known. These percentages are obtained from combustion data.

Elements present	C	H	N	O
Percentage	40.1	6.6	0.00	53.3
Relative atomic mass	12	1	14	16
Percentage/Relative atomic mass	3.34	6.60	0.00	3.33
Divide by smallest (3.33)	1	2	0	1

Therefore the **empirical** formula is CH_2O. The empirical formula contains the elements in the correct proportions, the actual formula can be an integer multiple of this. Percentages for the elements C, H and N are usually measured. Oxygen is not and an assumption is made that the remaining percentage after all the others are taken into account corresponds to that of oxygen. Percentages of other elements in a molecule can be measured, for example sulfur. The principle for working out the empirical formula is exactly the same in these cases.

1.5 Structural Isomers

There are many cases where molecules with the same molecular formula have different structural formulae. These are known as **structural isomers**.

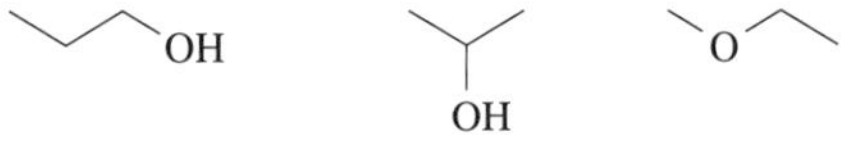

OH OH O

1.6 The Shape of Simple Molecules

The **shapes of simple organic molecules** are determined by the nature of the bonding between the atoms of the molecules.

1.7 Hybridisation

(See Gray 1973.)

Hybridisation was an early attempt at rationalising the shapes and bonding in molecules containing carbon (it was also applied to other atoms). Molecular orbital theory has since developed a more real description of bonding. However, organic chemists still use the hybridisation model of bonding as it is easy to understand and explains much of organic chemistry.

A carbon atom contains two electrons in its 1s orbital and two electrons in its 2s orbital and two electrons in two of its three 2p orbitals.

So normally the electronic configuration of carbon is described as $1s^2 2s^2 2p_x{}^1 2p_y{}^1 2p_z{}^0$. Although it requires energy to promote a 2s electron into the vacant 2p orbital, this energy can be regained if the 2s and three 2p orbitals, each now containing one electron, are mixed together to produce four new equivalent orbitals. This mixing process is termed **hybridisation**.

On promotion we obtain, $1s^2 2s^1 2p_x{}^1 2p_y{}^1 2p_z{}^1$, and on hybridisation we obtain 1s $(sp^3)^1(sp^3)^1(sp^3)^1(sp^3)^1$, where sp^3 symbolises the new hybridised orbital.

The sp^3 orbitals look like:

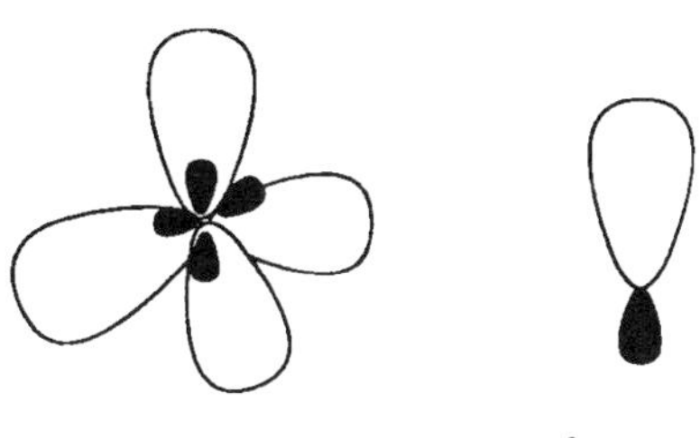

the four sp^3 orbitals a single sp^3 orbital

The four sp orbitals are arranged such that they point towards the corners of a tetrahedron.

This process can be repeated, but taking only two of the 2p orbitals and the 2s orbital. The mixing or hybridisation of these three atomic orbitals produces the three molecular sp^2 orbitals. These are arranged such that they point towards the corners of a triangle (they are all in the same plane). Furthermore, only the 2s and one 2p orbital can be taken and hybridised to give an sp orbital. These are arranged in a linear fashion. Both the sp^2 and the sp orbitals are drawn below.

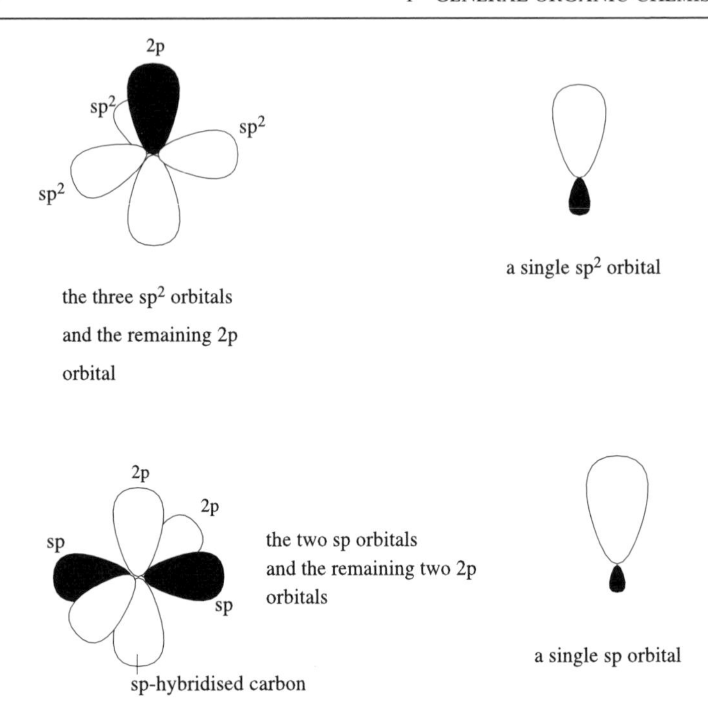

The sp^3, sp^2 and sp orbitals are the main orbitals used when discussing the bonding and shapes of organic molecules.

1.8 The Conformations and Shapes of Ethane and Butane

(Marples 1981 is worth browsing through.)

In ethane each carbon has four groups attached, also each carbon is sp^3 (tetrahedral, the angles between the C–H bonds are 109°) hybridised.

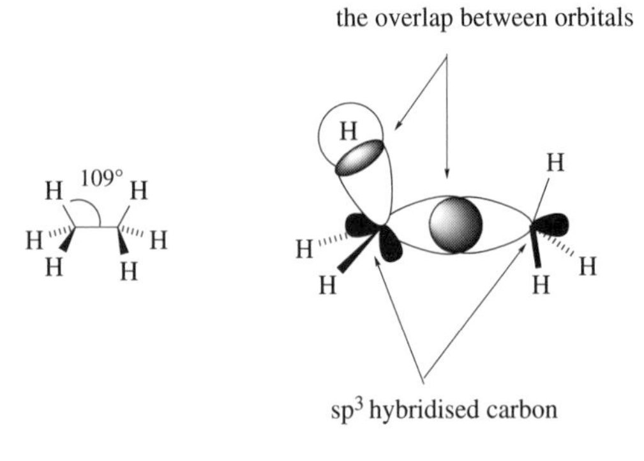

Sigma (σ) molecular orbitals, or sigma bonds, are cylindrically symmetrical about the interatomic axis, hence essentially "free" rotation is possible about the carbon–carbon bond.

free rotation about the C-C bond

The shapes of other functional groups and their roles in dictating the overall shape of a molecule will be dealt with as they arise in the text.

Ethane can thus be drawn in many different ways or **conformations**. The conformation of a molecule is the relationship of all the atoms in the molecule to one another in three dimensions. Ethane has an infinite number of conformations; however, the molecule has two **limiting** conformations (those conformations which have the maximum and minimum energies): the **eclipsed** and the **staggered** conformations. These can be shown as **Newman** projections:

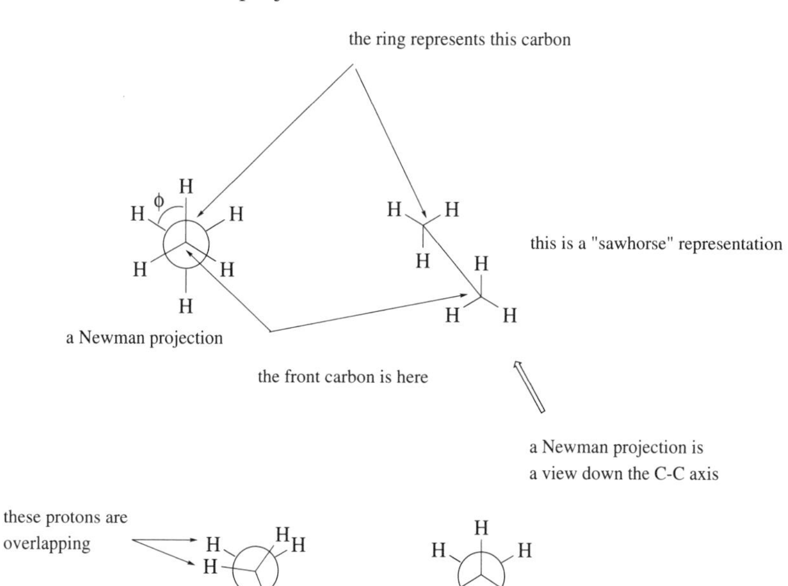

The angle ϕ is known as the angle of torsion. This angle is 0° for the eclipsed and 60° for the staggered conformation.

Torsional strain, see 1.13, is maximum for eclipsed and minimum for staggered. The energy of the molecule is highest when eclipsed and lowest when staggered.

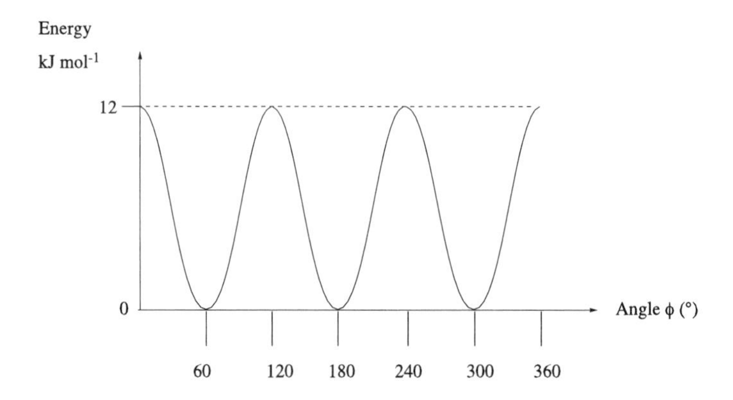

Rotation about a carbon–carbon single bond is not entirely free, but the energy barrier to rotation is very small, around 12 kJ mol^{-1}, and so rotation is very fast, approximately 10^{12} revolutions s^{-1}, at room temperature. Thus ethane acts as a single molecule (all molecules of ethane can be considered the same). The molecule spends most of its time in the staggered conformation and therefore the shape of ethane (approximately) is ethane in the staggered conformation.

With butane there are six limiting conformations:

overlapping

0°	60°	120°	180°	240°	300°
fully eclipsed	staggered (gauche)	partial eclipse	fully staggered (anti)	partial eclipse	staggered (gauche)

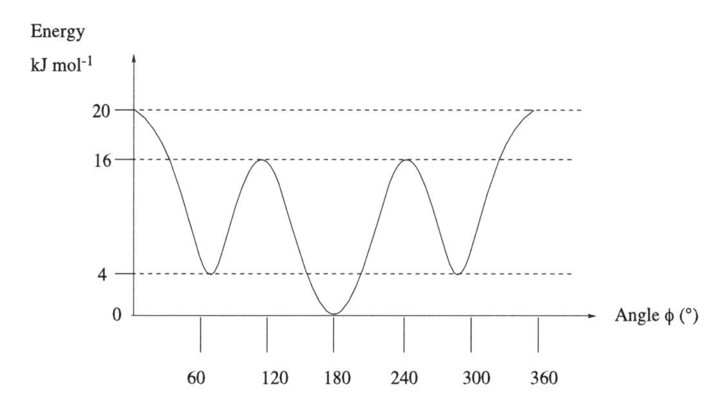

The energy barrier between the conformations is still small compared to the thermal energy of the molecule. To all intents and purpose the rotation is free. Most of the time the molecule is in the staggered (gauche) G and staggered (anti) A positions. There is a Boltzmann distribution between these two stable conformations. There is 70% A and 15% in each of the G conformations at 25°C.

1.9 The Conformations of Cyclohexane

Cyclohexane can exist in several possible conformations. These are: the **chair**, the **boat**, the **skew boat** and the **raft**. Of these possible conformations only the first three are of any importance. The raft is the worst possible arrangement for a cyclohexane ring, as the torsional strain, see 1.13, is extremely high.

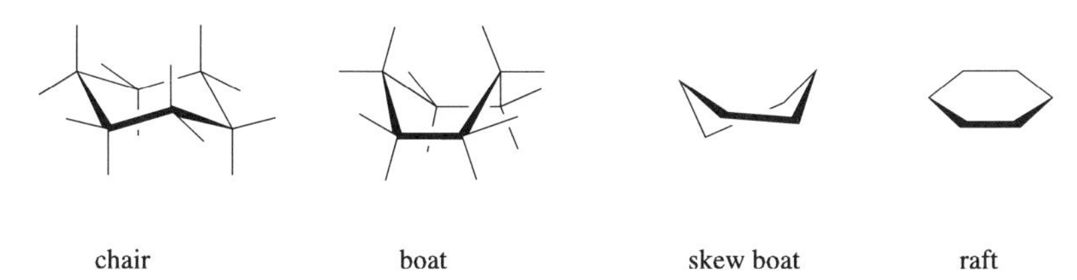

chair boat skew boat raft

The two minimum energy forms of cyclohexane are the chair and its "**ring-flipped**" conformer. These interconvert *via* the skew boat form.

There are two types of bond possible in the chair form: these are the **axial** and **equatorial** bonds. Cyclohexane assumes the chair shape to minimise internal strain. A view down any carbon–carbon bond in the chair form gives a Newman projection similar to "gauche" butane. There is little angle strain and all angles are approximately 111°.

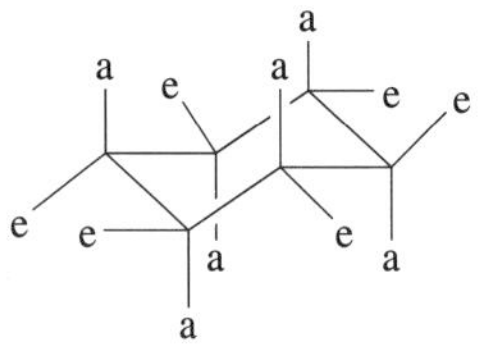

a = axial bonds

e = equatorial bonds

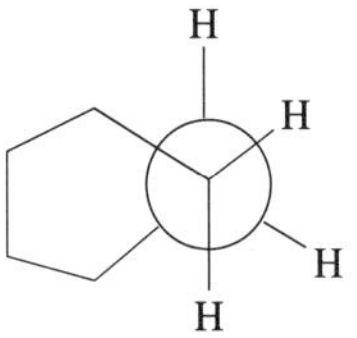

view down a C-C bond

The axial substituents in the chair form of cyclohexane can interact in what are known as **1,3 diaxial interactions**; this increases the torsional strain. There is also steric strain if the substituents are large. There is no steric interaction when hydrogen is in the axial position.

Cyclohexane can also exist in a boat form. In this conformation there are more than two types of bond (types 1 to 4 below). The boat form is of higher energy than the chair form. If the boat form twists slightly the energy of the molecule is reduced; this is called the skew boat.

When a cyclohexane has a substituent in the ring, it will usually adopt the chair conformation that places the substituent in the equatorial position.

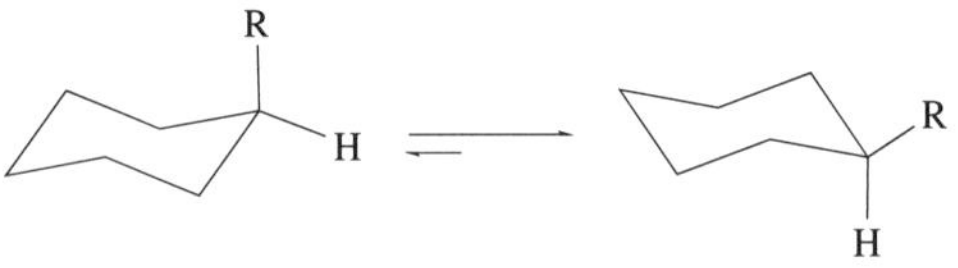

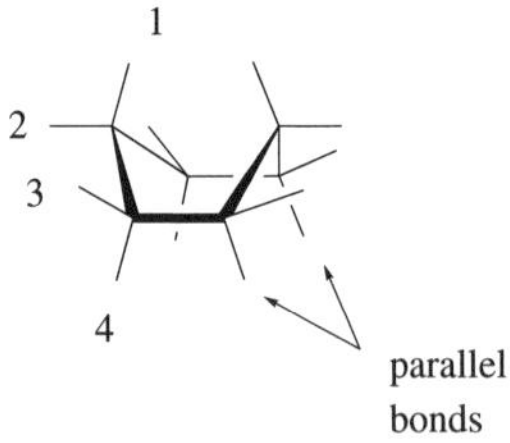

parallel bonds

This reduces the 1,3 diaxial interactions. The size, shape and bond length of the substituent determine how far the equilibrium lies to the right, below. Solvents can also play a large role, as they can increase the size, by coordination, of the substituent.

Some examples of how these effects operate are:

1. **Size:** *tert*-butyl groups will have more preference to be equatorial than methyl groups. Large groups on a cyclohexane ring are said to **conformationally anchor** the ring.
2. **Shape**: the cyano group, CN, is linear in shape. It therefore has almost no 1,3 diaxial interaction and hence has little preference for the axial or equatorial position.
3. **Bond length**: going from Cl to Br to I, the size increases, but so does the bond length. All three of these substituents have approximately the same preference for the equatorial position.
4. **Solvents**: carboxylic acids can be surrounded by solvent molecules; this makes them larger and hence they have a large preference for the equatorial position.

1.10 The Conformations of *trans* and *cis* Decalin

The bicyclic systems, *trans* and *cis* decalin are not conformational isomers, they are different compounds. *Trans* decalin is a rigid molecule and has no other conformation. It is called *trans* because the substituents (R) at the ring junction are *trans* to one another.

Cis decalin has two major conformers. One is the "ring-flipped" form of the other. Like cyclohexane, the major conformer will be the one which reduces the 1,3 diaxial interactions the most.

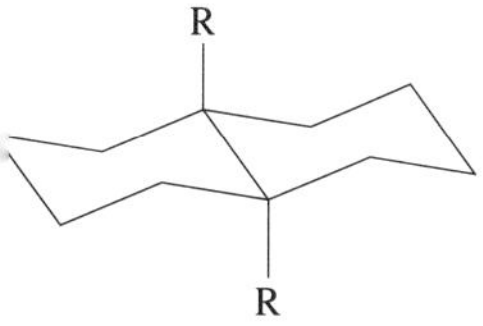

R R

trans decalin

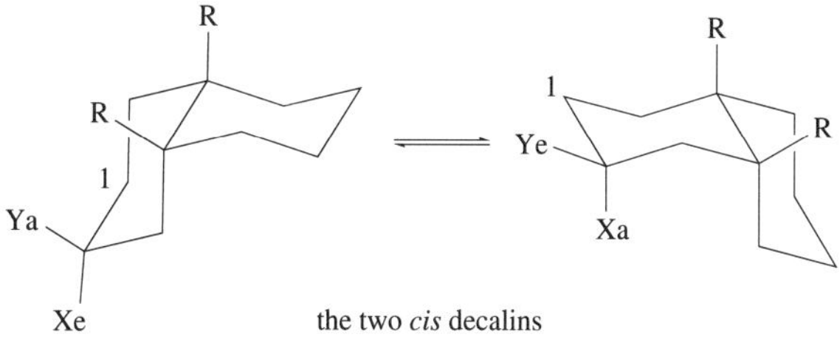

R R 1 Ya Xe R 1 Ye R Xa

the two *cis* decalins

1.11 The Conformations of Cyclohexene

When there is a double bond in the ring, the correct method for drawing the conformation is as shown below for cyclohexene. Cyclohexene also has a “ring-flipped” conformer. An epoxide in a cyclohexane ring should be drawn in a similar fashion.

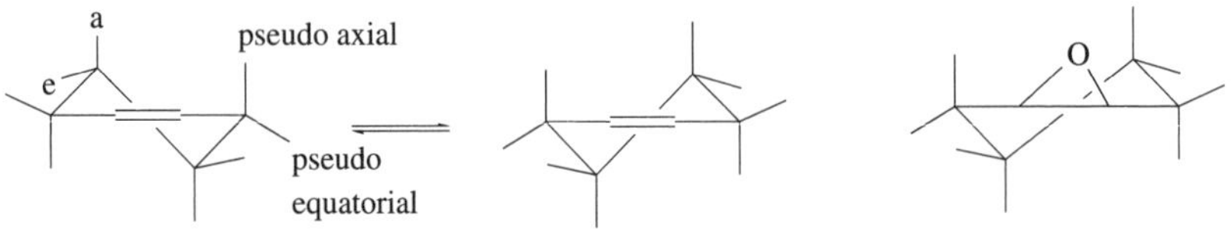

a
e
pseudo axial
pseudo
equatorial
O

the two conformers of cyclohexene

an epoxide in the cyclohexane ring

1.12 The Conformations of Other Ring Systems

The conformations of other ring systems are not as well studied as those of cyclohexanes.

Cyclopropane has high angle strain and high torsional strain. There is only one shape, cyclopropane is flat.

Cyclobutane is not flat, the angle strain is increased by distorting, but the torsional strain is reduced. There are 1,3 diaxial interactions.

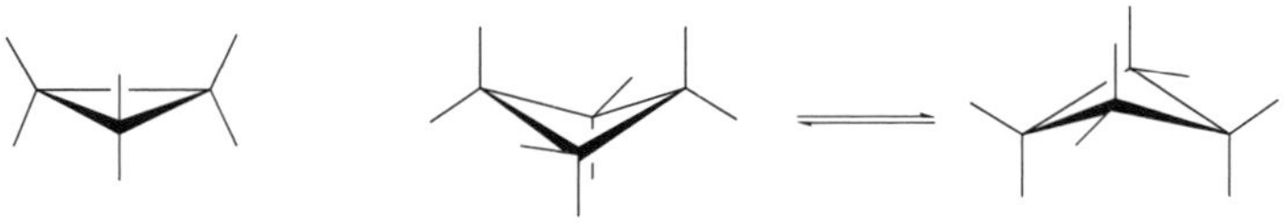

cyclopropane

the two conformers of cyclobutane

Cyclopentane has a “flap” which is continuously “rotating” about the ring; in other words, the carbon at the apex of the flap is continuously changing.

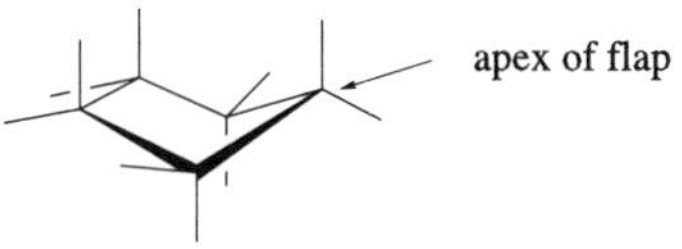

apex of flap

cyclopentane

Seven-membered rings exist in three main conformers, as do eight-membered rings.

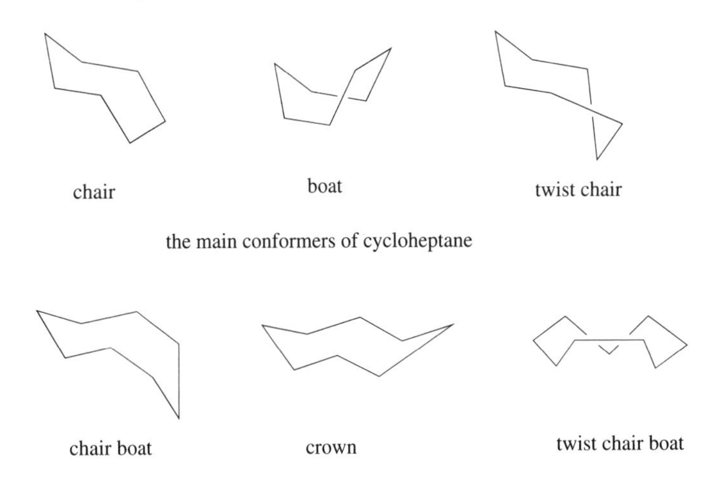

the main conformers of cycloheptane

the main conformers of cyclooctane

1.13 Torsional Strain

There are two components to torsional strain, steric or van der Waals repulsion and Pitzer strain. Torsional strain is the sum of these two forces. It is maximum for the eclipsed conformation and minimum for the staggered conformation.

1.14 Steric or van der Waals Repulsion

Steric or van der Waals repulsion between non-bonded atoms or groups occurs when they are in close proximity.

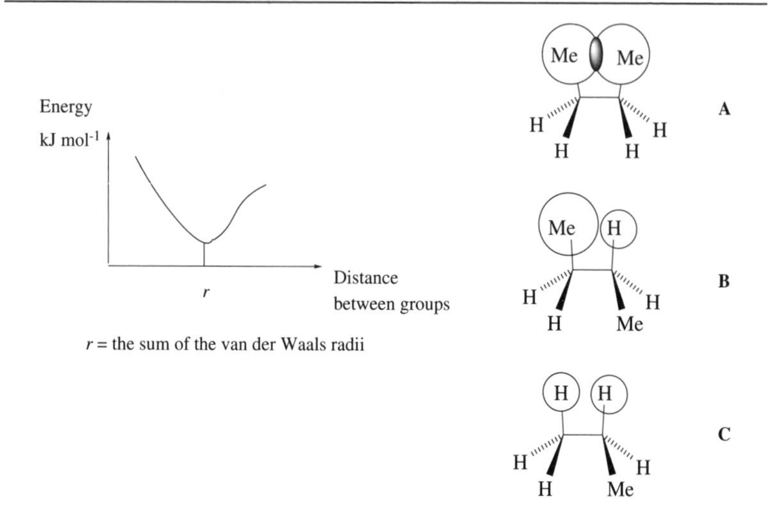

r = the sum of the van der Waals radii

When methyl eclipses methyl **A**, they are forced closer together than the sum of the van der Waals radii; therefore steric repulsion occurs. If two hydrogens eclipse **C** there is no steric repulsion because they are small. If a hydrogen and a methyl eclipse **B** the van der Waals radii just impinge on each other and therefore there is a small steric repulsion.

Steric repulsion is at its maximum for the eclipsed conformation and increases as the size of the atoms or groups increases.

1.15 Pitzer Strain

Pitzer strain is a repulsion between electron pairs in bonds.

repulsion

Me Me

H H

H H

Repulsion is a maximum when bonds are eclipsed. Pitzer strain is approximately the same for all groups of a similar nature.

1.16 Stereoisomerism

(Some books deal specifically with chirality. These often have a chapter on the "Description of stereochemistry". These sections detail stereochemistry much more than the average text, see Aitkin and Kilényi 1994.)

Stereoisomers have the same molecular and structural formulae but differ in the way the atoms are arranged in space. The **configuration** of a molecule is the arrangement of atoms in space that characterises that particular **stereoisomer**.
Stereoisomerism can arise from **restricted rotation** about double bonds or from **chirality**.

1.17 Restricted Rotation about Double Bonds

The molecular formula C_4H_8 gives rise to three **structural isomers**:

a) $MeCH_2CH{=}CH_2$ b) $MeCH{=}CHMe$ c) $Me_2C{=}CH_2$

b) has two forms:

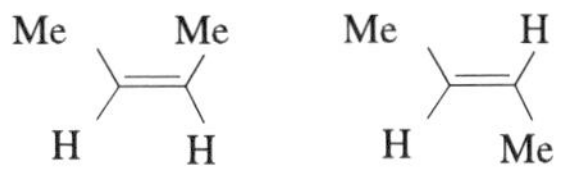

Me Me Me H

H H H Me

cis *trans*

When the shape of the molecules is considered there are really four isomers because (b) exists in two forms. These are known as the *cis* and *trans* isomers (there is another nomenclature system which is considered in 1.25).

Consider the ethene molecule, C_2H_4, each carbon of which has only three other groups attached to it. The carbon has sp^2 hybridisation.

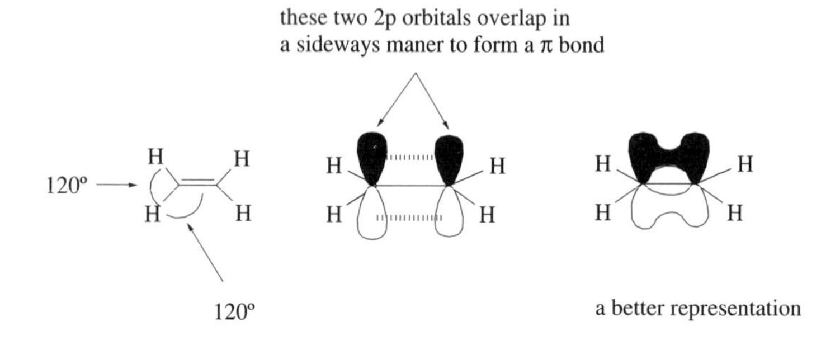

this means that these lobes of the orbital are overlapping

Even though a pi (π) bond ($275\,\mathrm{kJ\,mol^{-1}}$) is weaker than a σ bond ($400\,\mathrm{kJ\,mol^{-1}}$) (because sideways overlap of orbitals is less effective than end-on overlap) there is still restricted rotation about the carbon–carbon double bond (the term used to describe a bond between two atoms consisting of a sigma and a pi bond) at room temperature. The energy barrier to rotation is $275\,\mathrm{kJ\,mol^{-1}}$ so there is only interconversion between the two isomers at elevated temperatures. The *cis* and *trans* isomers can be considered as two distinct molecules at normal temperatures. In the case of but-2-ene and other alkenes, the *cis* and the *trans* isomers have different physical and chemical properties.

1.18 Occurance of Geometrical Isomerism

For an alkene:

a c / b d   a d / b c

geometrical isomerism exists if a $\neq$ b and c $\neq$ d.

If there is more than one carbon–carbon double bond, there is a possibility of more than two geometrical isomers:

MeCH=CHCH=CHMe

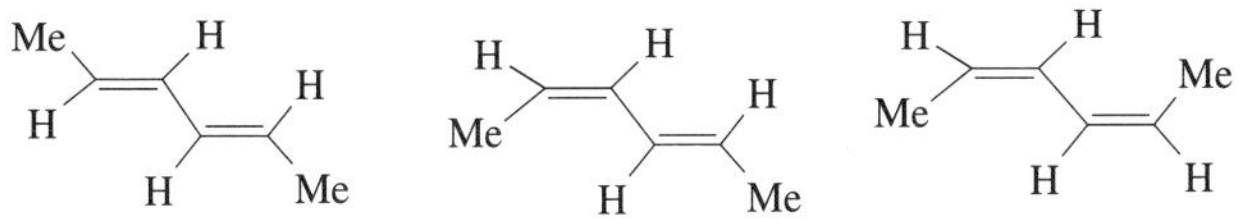

trans trans *cis trans* *cis cis*

2,4-Hexadiene has three geometrical isomers and 2,4-heptadiene has four. For a molecule with *n* double bonds there is a maximum possibility of 2^n geometrical isomers.

Other double bonds, for example C=N and N=N, can also give rise to geometrical isomers.

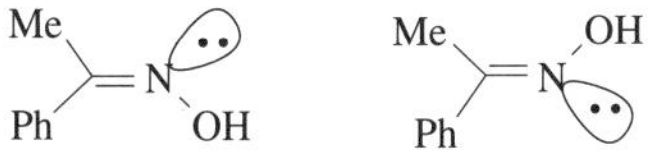

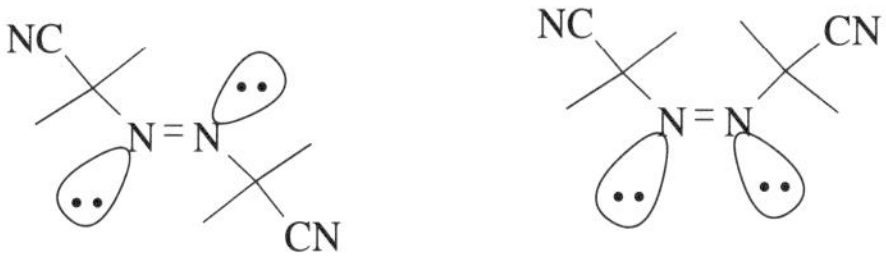

1.19 Chirality

(See March 1992 and Aitken and Kilényi 1994.)

In order to understand the concept of chirality it is vitally important that an ability to think in three dimensions is developed. An invaluable aid is a set of cheap "ball and stick" models.

1.20 Polarimetry and Optical Activity

Plane-polarised light (ppl) vibrates in one plane only. Compounds exist that rotate the plane of ppl if it is passed through them. These compounds are said to be **optically active**. If the compound shows no rotation it is said to be **optically inactive**.

A polarimeter:

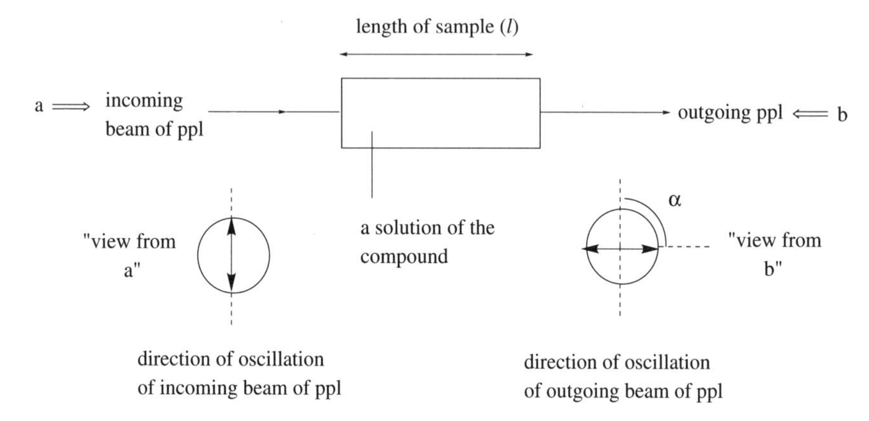

The angle α through which the ppl has been rotated can be measured. A sample can give a clockwise (+ or dextro (*d*)) rotation or an anticlockwise (− or laevo (*l*)) rotation.

The angle α through which the ppl is rotated is dependent on the concentration of the sample and the length of the sample (operator variables).

The specific rotation $[\alpha]$ is given by:

$$[\alpha] = \alpha/c.\,l$$

where α is the measured rotation, c is the concentration (in $\mathrm{gmL^{-1}}$) and l is the length in dm.

For example: menthol has a measured rotation a of $-4.1°$ for a solution of 0.84 g in 10 mL of solution. The cell was 100 mm long. So $c = 0.084$ and $l = 1$. Therefore $[\alpha] = -4.1/0.084 = -48.8°$.

The angle also depends on the wavelength and temperature. In most cases the sodium D wavelength ($\lambda = 589$ nm) is used. The wavelength λ used is usually quoted.

$$[\alpha]\lambda = \alpha/c.\,l$$

or

$$[\alpha]_D = \alpha/c.\,l \text{ if the sodium D wavelength is used.}$$

These equations are for solutions. If a pure sample is used, then

$[\alpha]_D = \alpha/l.d$ where d is the density.

The temperature, T, at which the rotation was measured at is given as a superscript:

$$[\alpha]_D{}^T.$$

1.21 Enantiomers

The molecules of some compounds are optically active, or, as stated above, they rotate ppl. A compound such as CH_3CHCl_2 is optically inactive but the compound $CH_3CHClBr$ is optically active. Why should this be so?

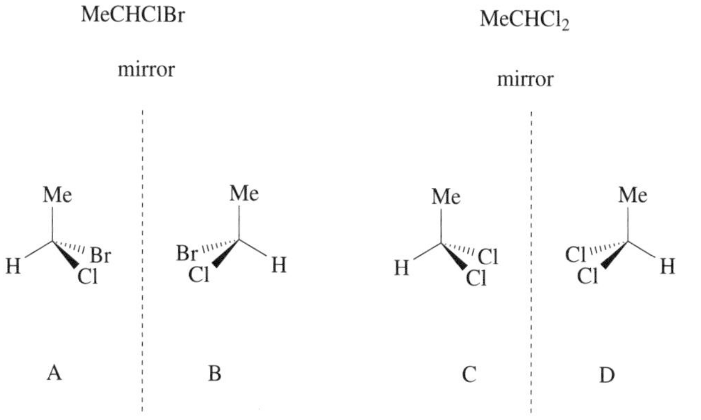

MeCHClBr

mirror

A B

two non-superimposable mirror images or stereoisomers

these are therefore enantiomers

$MeCHCl_2$

mirror

C D

two superimposable mirror images

these are therefore not enantiomers

The molecule $CH_3CHClBr$ displays optical activity if pure A or pure B is used (or if either A or B is in excess over the other. Note that the two mirror images A and B cannot be superimposed on one another (construct models). However, the mirror images C and D of CH_3CHCl_2 can be superimposed on each other and there is no optical activity.

Another way of determining whether a molecule is chiral or not is to investigate its symmetry. If there is a plane of symmetry the molecule cannot be chiral.

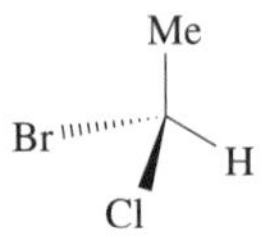

Me
Br H
Cl

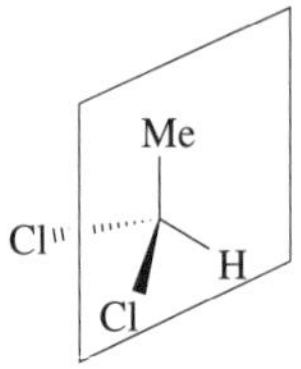

Me
Cl H
Cl

no plane of symmetry: chiral

plane of symmetry: achiral

Some points to note:

1. **Enantiomers** are stereoisomers such that one is the **non-superimposable** mirror image of the other (as in A and B above).
2. The stereoisomers A and B are an enantiomeric pair.
3. A molecule is said to be chiral if its mirror image is non-superimposable.
4. A compound whose molecules are chiral (have non-superimposable mirror images) exists as a pair of optically active enantiomers
5. A compound whose molecules are achiral (have superimposable mirror images, as in C and D above) does not exist as enantiomers and cannot be optically active.
6. A **racemate** (or **racemic mixture**) is an equal mixture (1:1) of two enantiomers. A racemic mixture will not rotate ppl.
7. The atom at the centre of the molecule with the four groups attached to it (in the case of A or B this is a carbon–H, Me, Cl and Br being the four attached groups) is known as the **chiral centre**.
8. A chiral centre is also called a **stereogenic centre**.
9. Compounds whose molecules have only one chiral centre, $CH_3CHClBr$, $HOCH_2CH(OH)CHO$, always exist as a pair of optically active enantiomers. Two of the groups may differ at some point remote from the chiral centre.

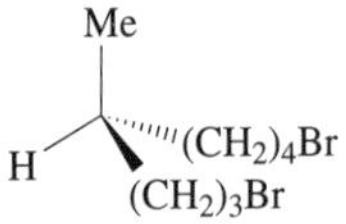

Me
H $(CH_2)_4Br$
$(CH_2)_3Br$

10. The chiral centre does not need to be carbon.

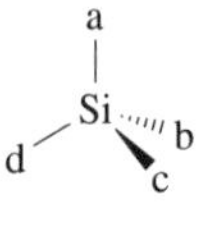

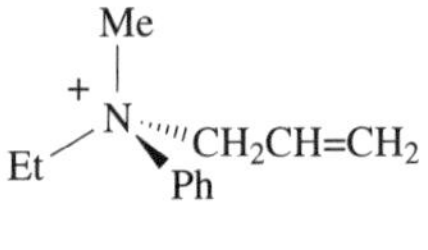

a
d Si b
c

Me
+
Et N $CH_2CH{=}CH_2$
Ph

1.22 Optical Purity

This is an indication of the excess of one enantiomer over another in a mixture:

$(+)$ cholesterol $[\alpha] = +39.0^\circ$,
$(-)$ cholesterol $[\alpha] = -39.0^\circ$.

A mixture containing 60% of the $(+)$ and 40% of the $(-)$ has $[\alpha] = +(39.0 \times 20)/100 = 7.8^\circ$. The mixture has a 20% **enantiomeric excess** (ee) so we observe 20% optical purity.

1.23 The Cahn–Ingold–Prelog (CIP) Rules (or the *R* and *S* Nomenclature System)

A system for describing each enantiomer of an enantiomeric pair has been devised. The individual enantiomers are named on the *R* and *S* system (or the Cahn–Ingold–Prelog (CIP) system).

The *R* and *S* system (CIP) sequence rules are laid out as follows:

1. The substituents attached to the chiral centre are prioritised according to the atomic number that the atom attached directly to the chiral centre has. You can label the atoms/groups as you wish; 1 (highest) to 4 (lowest) is used here.

   Lowest priority (4) H < C < N < O < Cl < P < Br **Highest priority (1)**

2. If two or more atoms attached directly to the chiral centre are the same, then it is the next atom along in the substituent that determines the priority; if these are the same it is the third and so on until one group has priority.

   $BrCH_2$- has a higher priority than $HOCH_2$-

3. Isotopes are treated in the same manner as in rule 1.

   Hydrogen (H) < Deuterium (D) < Tritium (T)

4. Multiple bonds are treated as shown below.

note R^1 can be RNH-, R_2N-, RO-, alkyl or H (*i.e.* any carbonyl function)

(partial expansion only)

5. A peculiarity of the system states that all atoms have a valency of four. Those atoms with a valency of less than four are given a phantom atom (PM) which has the lowest ranking of all. (Do not worry too much about this rule, just remember protonated nitrogen is higher priority than non-protonated nitrogen.)

$^{+}NHMe_2 > NMe_2$ because it is really $(PM)NMe_2$

Once the four groups are prioritised, view the molecule such that the lowest priority group is pointing away from yourself. Look at the three remaining groups facing you.

When the group priorities follow a clockwise direction the chiral centre is labelled *R*.

When the group priorities follow an anticlockwise direction the chiral centre is labelled *S*.

so the labelling is

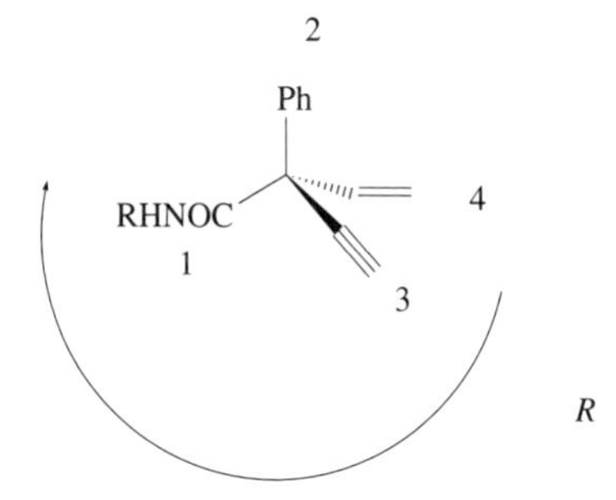

1.24 *R* and *S* from Fischer Projections

The same sequence rules as in the CIP system are used. Work out the priorities for the four ligands attached to the chiral centre. View the structure from the side opposite the ligand of lowest priority and consider the spatial arrangement of the remaining ligands.

When the group priorities follow a clockwise direction the chiral centre is labelled *R*.
When the group priorities follow an anticlockwise direction the chiral centre is labelled *S*.

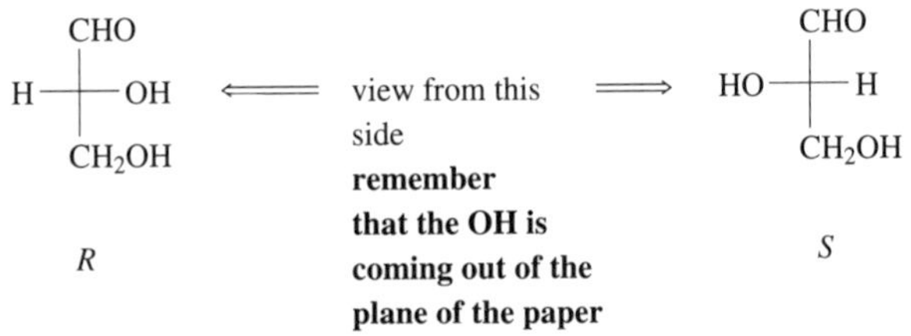

CHO
H—+—OH ⟸ view from this side **remember that the OH is coming out of the plane of the paper** ⟹ HO—+—H
CH_2OH

R *S*

1.25 Extension of CIP Rules to Double Bonds

An extension of the CIP system has been applied to double bonds. It is difficult to use the *cis* and *trans* description of a double bond if all the substituents are different. There is an alternative method based on the CIP rules. The four groups on the two carbons forming the double bond are ranked in exactly the same maner as in the *R* and *S* system. Then that isomer with the two highest-ranking groups on the same side is labelled *Z* (from the German *zusammen*, together) and the isomer with the two highest ranking groups on opposite sides is labelled *E* (from the German *entgegen*, opposite). Some examples are shown below.

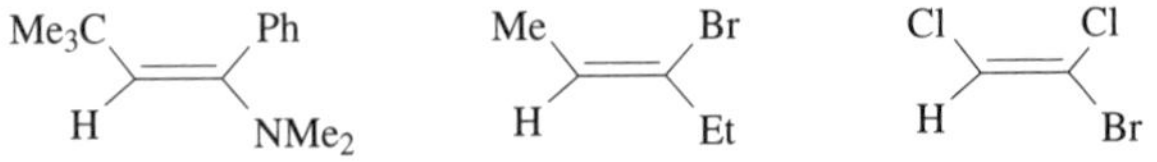

Me_3C(H)C=C(Ph)(NMe_2) — *E*

Me(H)C=C(Br)(Et) — *Z*

Cl(H)C=C(Cl)(Br) — *E*

1.26 Notes on the D and L System

The D and L system of denoting chirality arose from an arbitrary decision to call one enantiomer of glyceraldehyde D (for the one with the positive (+) rotation) and the other L (for the one with the negative (−) rotation). A monosaccharide which has its highest-numbered chiral carbon with the same configuration as D−(+)−glyceraldehyde is designated D. One in which it has the same configuration as L−(−)−glyceraldehyde is labelled L. The system was found to have many flaws and has generally dropped out of use.

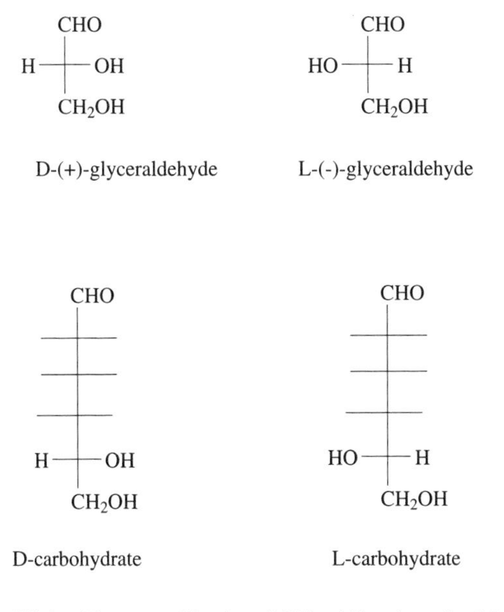

(this is with any combination of OH and H at the ends of the unoccupied horizontal bonds)

1.27 Diastereomers

Many compounds have more than one chiral centre. If there are *n* chiral centres then there are possibly 2^n stereoisomers. There are less than 2^n stereoisomers if there is some element of symmetry present.

The molecule $CH_3CHClCHBrCH_3$ has two chiral centres and therefore four stereoisomers or two pairs of enantiomers:

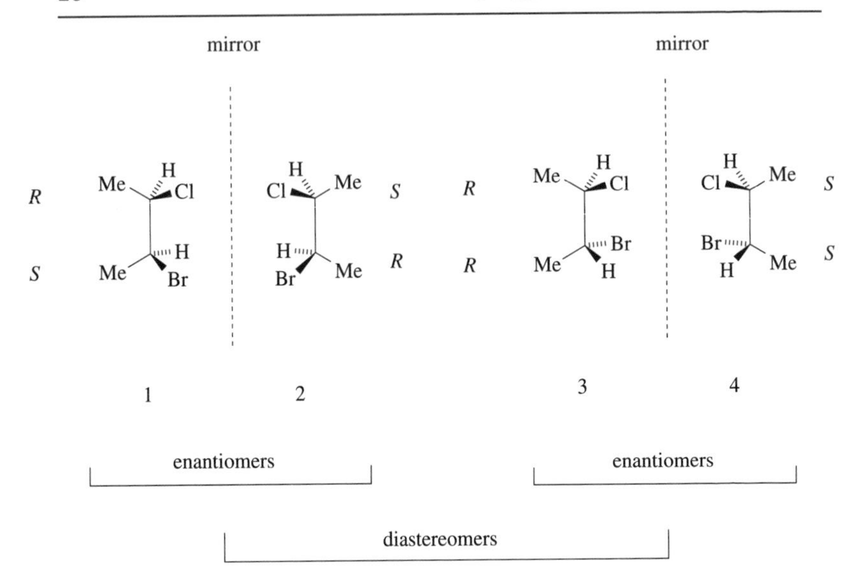

1 is the enantiomer of 2, a diastereomer of 3 and 4.
2 is the enantiomer of 1, a diastereomer of 3 and 4.
3 is an enantiomer of 4, a diastereomer of 1 and 2.
4 is the enantiomer of 3, a diastereomer of 1 and 2.

Diastereomers are any stereoisomers that are not enantiomers. Diastereomers have different chemical properties and different physical properties.

For example, the melting points, solubilities, infrared spectra and magnitude of [α] will differ. It is possible to separate diastereomers by normal methods, such as distillation, crystallisation and chromatography. This is not possible with enantiomers.

$$\text{1, 2, 3 and 4} \xrightarrow[\text{of separation}]{\text{normal methods}} \begin{cases} \text{1 and 2} \\ \text{3 and 4} \end{cases}$$

1.28 Meso Compounds

If an element of symmetry is present then there are less than 2^n stereoisomers. For example, $CH_3CHBrCHBrCH_3$ has only three stereoisomers.

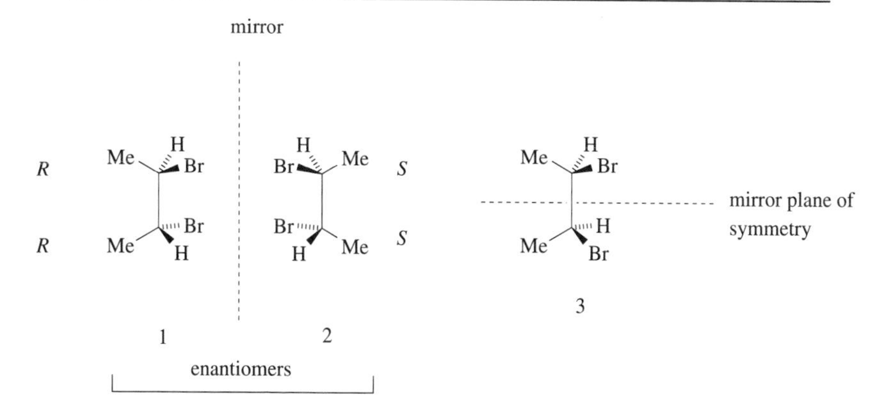

Molecules 1 and 2 are stereoisomers, 1 is the enantiomer of 2. Molecule 3 has an internal plane of symmetry, it is optically inactive even with two chiral centres. Molecule 3 is called a **meso compound**. It is a diastereomer of 1 and 2.

A simplistic view as to why 3 is optically inactive is that ppl rotated one way by the top half of the molecule is rotated the other way by the bottom half (internal compensation).

Cyclic compounds can also be optically active (* = chiral centre), 1,2-dibromocyclobutane can also exist in a meso form;

1.29 Optical Activity with No Chiral Centre

(See March 1992.)

Some molecules are optically active even with no chiral centre (central chirality). Substituted allenes are such compounds.

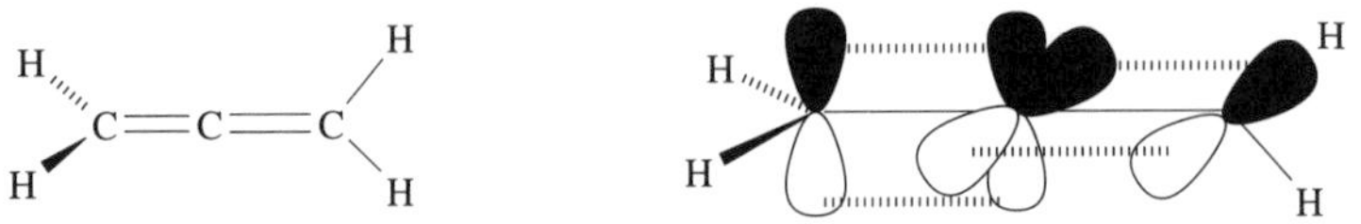

achiral

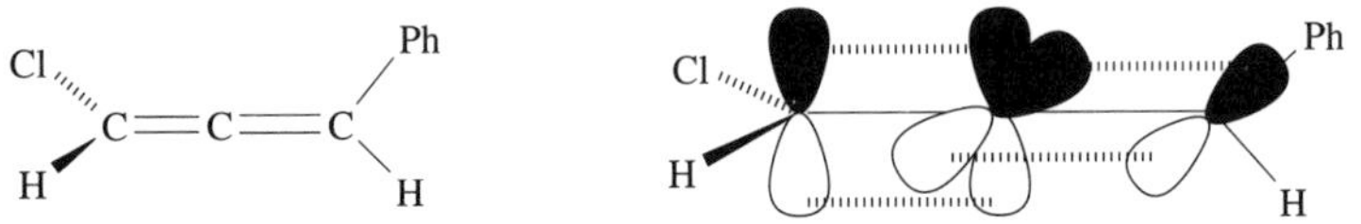

chiral

They are chiral when for a general allene abC=C=Ccd, a ≠ b and c ≠ d. So MeCH=C=CHMe is a chiral compound and exists as a pair of enantiomers (this is axial chirality). There are other types of chirality such as helical and planar.

1.30 Resolution of a Racemic Mixture into Enantiomers

It is usually possible to resolve (separate) the two enantiomers in a racemic mixture. This is normally done by reversible conversion into diastereomers using another pure enantiomer as a reagent.

racemate (+) and (-) ⟶ enantiomer (+) and enantiomer (-)

compound A (+) and (-) —compound B / a single enantiomer (+) (this is the resolving agent)⟶ diastereomer A (+) B (+) and diastereomer A (-) B (+)

The resulting diastereomers can be separated using their differences in physical properties. Once separated the diastereomers are converted back to the enantiomers.

diastereomer A (+) B (+) —chemical separation of A (+) and B (+)→ enantiomer A (+) and enantiomer B (+)

diastereomer A (-) B (+) —chemical separation of A (-) and B (+)→ enantiomer A (-) and enantiomer B (+)

1.31 Reactions at Chiral Centres

There are three outcomes for reactions at a chiral centre. Consider the replacement of a group X by another group Y.

Et X H Me —Y→ Et Y H Me

single enantiomer

if this is the only compound obtained then the process is known as **stereospecific retention of configuration**

—Y→ Et H Me Y

if this is the only compound obtained then the process is known as **stereospecific inversion of configuration**

—Y→ Et Y H Me and Et H Me Y

if a mixture of the two above situations occurs then the reaction process is known as a **non-stereospecific reaction (or racemisation)**

These reactions occur through mechanisms denoted by S_N1 and S_N2, the S_N1 and S_N2 reactions are discussed in Chapter 8.

1.32 Substituent Effects in Molecules

The substituents in a molecule will greatly affect the reactivity and chemical behaviour of the molecule.

There are essentially two different effects. There is the inductive (**I**) effect and the resonance (**R**) effect.

The substituent X is going to be either electropositive or electronegative. If X is electronegative then electrons are pulled towards X, a $-$**I** effect ($\delta+$ and $\delta-$ mean partial charge, $\delta\delta+$ and $\delta\delta-$ mean a smaller partial charge still).

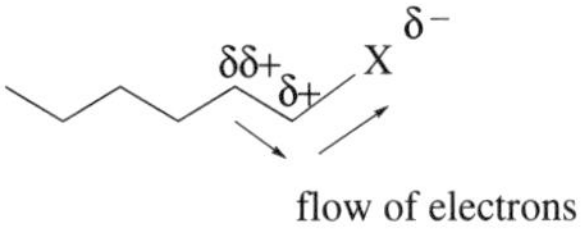

$\delta\delta+$ $\delta+$ X $\delta-$

flow of electrons

$\delta\delta$ means that there is less effect of X on this carbon

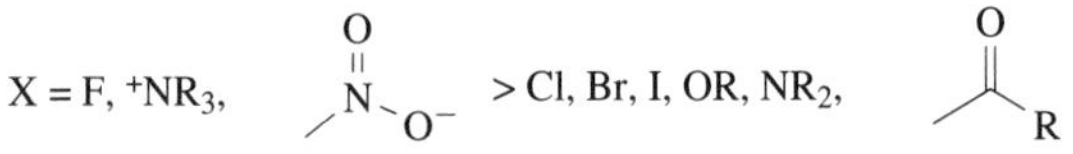

X = F, $^+NR_3$, $-N(=O)O^-$ > Cl, Br, I, OR, NR_2, $-C(=O)R$

If X is electropositive then X exerts an electron-releasing effect, +**I**.

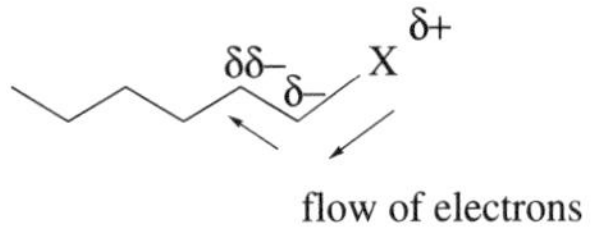

$\delta\delta-$ $\delta-$ X $\delta+$

flow of electrons

X = Li, Na, K, MgX and alkyl groups relative to hydrogen

The concept and effect of resonance on a system are dealt with in Section 1.39.

1.33 Homolysis of Bonds

A pair of bonding electrons are shared between the fragments when the bond breaks. The fragments are **(free) radicals**. For the movement of one electron a half-headed arrow (or "fish hook") is used.

$$A\text{—}B \longrightarrow A\bullet + B\bullet$$

= movement of one electron

1.34 Heterolysis of Bonds

One fragment takes both bonding electrons when the bond breaks. The fragments are ions.

$$A\text{—}B \longrightarrow A^{+} + B^{-}$$

= movement of two electrons

The movement of electrons in a reaction is denoted by an arrow. For movement of two electrons a full-headed arrow is used. The use of arrows is a "book-keeping" procedure. The arrows always show where **electrons** (or **electron density**) are going. For several arrows in a row the head must generally follow the tails, see Section 1.39.

1.35 The Structure of Carbocations, Carbon (Free) Radicals and Carbanions

For a compound R_3C–X when the C–X bond breaks we can obtain:
a **carbocation (carbonium)** ion (C^+),

empty orbital

X

$- X^-$

R^3 R^1 R^2 R^3 R^1 R^2

$R^1R^2R^3C^+$

there is an sp^2 hybridised central carbon with 6 bonding electrons, there is a vacant p orbital and the molecule is planar

there is an overall positive charge

a **carbon (free) radical** (C•),

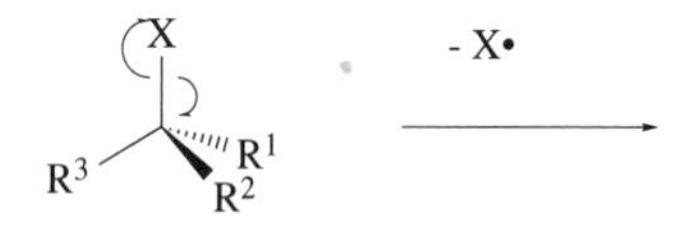

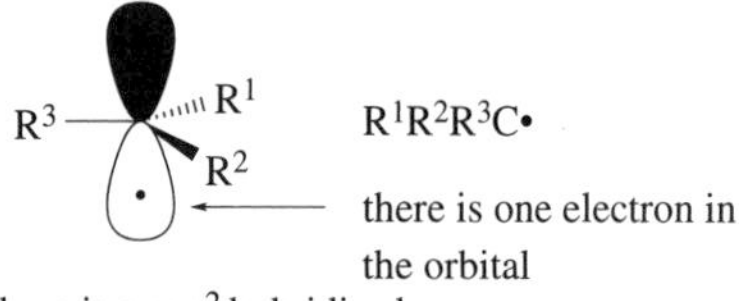

there is an sp^2 hybridised central carbon with 7 bonding electrons, there is a single electron in the p orbital and the molecule is planar

there is no overall charge

or a **carbanion** (C^-),

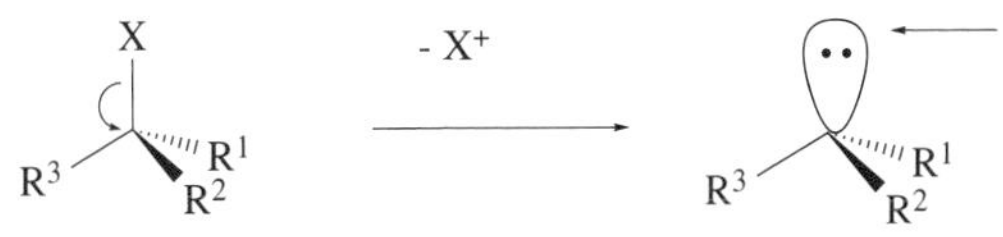

X

R3 R1 R2 — - X+ → R3 R1 R2

there are two electrons in the orbital

$R^1R^2R^3C^-$

there is an sp^3 hybridised central carbon with 8 bonding electrons, two of these electrons are in the sp^3 orbital and the molecule is tetrahedral

there is an overall negative charge

All of these three species are high in energy. They are usually very reactive and have lifetimes of nanoseconds only. They can be primary (1°), secondary (2°), or tertiary (3°) species in nature.

A primary carbon is bonded to only one other and three hydrogens. A secondary carbon is bonded to two others and two hydrogens. A tertiary carbon is bonded to three others and one hydrogen.

When one of the hydrogens is "replaced" by a single electron (radical), positive charge (carbocation) or negative charge (carbanion), the new species can also be termed primary, secondary or tertiary depending on the number of attached carbons. For example, in the case of radicals:

primary	$CH_3CH_2\bullet$
secondary	$(CH_3)_2CH\bullet$
tertiary	$(CH_3)_3C\bullet$

1.36 Nucleophiles

A nucleophile is any species which wants to donate electrons or electron density. It is important to realise that good nucleophiles are not necessarily good bases although the two often run parallel. "Nucleophile" is a kinetic term and "base" is a thermodynamic term. A powerful nucleophile is one which reacts rapidly with an electrophile (small energy of activation) and a strong base is one whose reaction with an acid reaches an equilibrium well to the right (large negative ΔG). Note that **negative charge is not in itself able to give rise to nucleophilicity**.

$H\bar{O}$, $\bar{C}N$, $\bar{O}Ac$, R^-

$\ddot{O}$ $\ddot{S}$ $\ddot{N}$

$R^{\delta-}—Li^{\delta+}$ $R^{\delta-}—MgX^{\delta+}$

1.37 Electrophiles

An electrophile is any species which wants to accept electrons or electron density. It is important to realise that good electrophiles are not necessarily good acids. "Electrophile" is a kinetic term, whilst "acid" is a thermodynamic term. Note that **a positive charge is neither necessary nor sufficient to confer electrophilicity upon a species**.

H^+, metal$^+$, NO_2^+

$AlCl_3$, BF_3

$HO^{\delta-}—Br^{\delta+}$ ← electrophilic part

1.38 Notes on Describing Reaction Mechanisms or "Arrow Pushing"

(The best method of learning this technique is practice, see Warren 1974, Warren 1978, Simpson 1994, Edenborough 1994 and Scudder 1992.)

All reactions have a **mechanism** through which the reactants are converted into products. For a few types of reaction the mechanism is known in detail. For the majority of reactions there are points about the mechanism which have not been fully explored, although usually something is known. There are a few reactions where the complexity of the reaction has so far defied all attempts at explanation. However, whenever a reaction is discussed, its mechanism should also be noted, or at least an attempt should be made at understanding what is really happening. Mechanisms, on the whole, are depicted in terms of electrons moving from one atom (or bond) to another, indicated by an arrow. Throughout this book the mechanisms for the more important reactions are given.

1.39 Resonance in Molecules

Resonance is a term used in the **valence bond** description of systems used to account for their higher stability in comparison to other systems. For a given structure it is often possible to distribute any π-bonding electrons or lone pairs of electrons in more than one way, at the same time maintaining the integrity of the structure.

A double-headed arrow is used to denote resonance. The two or more forms are known as **canonical forms** (or sometimes as **resonance forms**) and the molecule, a single species, is known as a **resonance hybrid**. It is important to note that only the **hybrid** form exists, the **canonical forms have no existence at all**. Beware of the term resonance: it implies that the molecule oscillates between the two forms, but this is not the case. See 3.34 for an example of resonance and the comparison of resonance and delocalisation.

1.40 Delocalisation in Molecules

Delocalisation is a molecular orbital term and describes stabilisation of systems by the spreading out or **delocalisation** of unpaired electrons or lone pairs of electrons over an extended π orbital system.

Delocalisation and resonance are different approaches to the problem of attributing stabilisation to certain systems: they often use the same diagrams to describe what is going on and often blur into one another. See 3.34 for an example of delocalisation and the comparison of resonance and delocalisation.

1.41 Formal Charge in Molecules

The formal charge is normally shown on each atom. For this purpose an atom is considered to "own" all unshared electrons and one-half of the electrons in covalent bonds. The sum of the electrons that "belong" to an atom is compared to the number "contributed" by the atom. An excess belonging to the atom results in a negative charge, and a deficiency results in a positive charge. The total of the formal charges on all the atoms equals the charge on the whole molecule or ion.

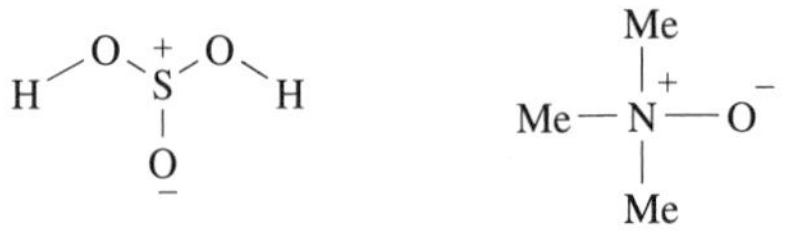

H—O—S⁺(—O⁻)—O—H $Me_3N^+—O^-$

1.42 Multistep Synthesis

In the rest of this book the organic compounds are effectively classified in terms of their **functional groups**; this classification allows an enormous amount of information to be organised into relatively few categories. Functional groups are typified by their chemistry and a characteristic property of one functional group is often a method of preparation for another. For example, a characteristic property of alcohols is that they react with mineral acids to give alkenes; hence alkenes can be prepared from alcohols.

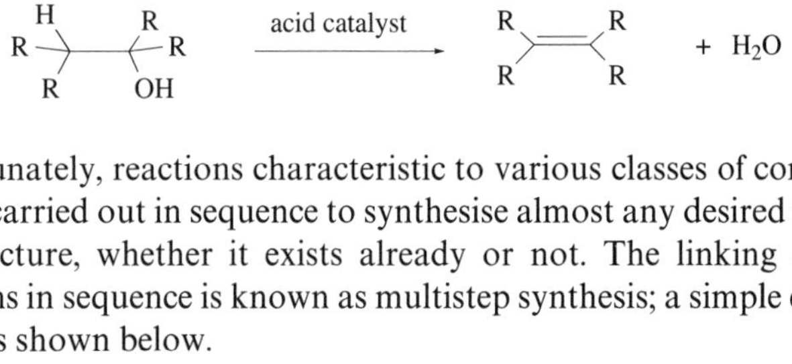

$$R_2CH–CR_2OH \xrightarrow{\text{acid catalyst}} R_2C{=}CR_2 + H_2O$$

Fortunately, reactions characteristic to various classes of compound can be carried out in sequence to synthesise almost any desired molecular structure, whether it exists already or not. The linking of these reactions in sequence is known as multistep synthesis; a simple example of this is shown below.

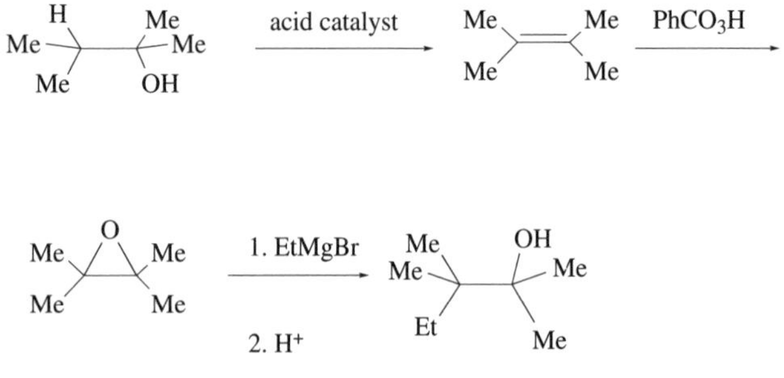

$$Me_2CH–CMe_2OH \xrightarrow{\text{acid catalyst}} Me_2C{=}CMe_2 \xrightarrow{PhCO_3H}$$

$$\text{Me}_2\text{C(–O–)CMe}_2 \xrightarrow[\text{2. } H^+]{\text{1. EtMgBr}} Me_2C(Et)–CMe_2OH$$

1.43 Retrosynthetic Analysis

Retrosynthetic analysis presents an analytical approach to the design of syntheses for molecules which we wish to make. It is the prelude to multistep synthesis. Retrosynthesis is the process of breaking down a target molecule into available materials by functional group interconversion and disconection. Functional group interconversion is the process of converting one functional group into another by substitution, addition, elimination, oxidation, reduction, rearrangement, *etc.* The term disconnection describes the reverse operation to a reaction. It is the imagined cleavage of a bond to break up the molecule into possible starting materials. Disconnections or functional group interconversions are denoted by the arrow shown in the simple example below.

An idealised fragment is known as a synthon. This is usually a cation or anion resulting from a disconnection. It may or may not be an intermediate in the reaction. A reagent is the compound used in practice for a synthon.

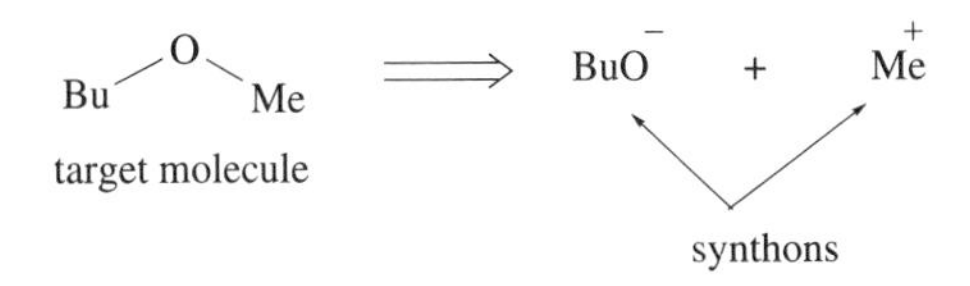

reagents in practice are BuOH/EtONa and MeI

It is not possible in a book of this kind to go through retrosynthetic analysis in detail. Readers are strongly recommended to consult Warren (1978) and Corey and Cheng (1989).

1.44 Relevant Questions

1, 11, 13, 24, 30, 34, 39, 43(a), 49(i), 51, 63, 85, 95(b), 98(b), 99, 102, 117(b), 124, 125, 130, 131(b), 137(e), 138, 142, 145, 152, 153, 156, 159, 165, 168.

2 ALKANES

2.1 Alkanes in General

Alkanes are saturated hydrocarbons. There are two types, the acyclic and the cyclic alkanes. Acyclic alkanes have the general formula C_nH_{2n+2}; when n (an integer) = 1, 2 or 3 there is only one possible structure and when $n \geq 4$ several structural isomers are possible.

For the molecular formula C_5H_{12} three isomers exist:

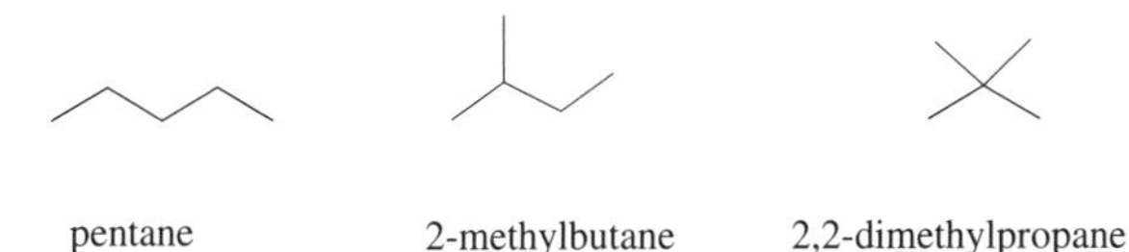

pentane 2-methylbutane 2,2-dimethylpropane

For the alkane C_7H_{16} there are nine possible structural isomers. For C_9H_{20} there are thirty-five possible structural isomers.

Monocyclic alkanes have a general formula C_nH_{2n}.

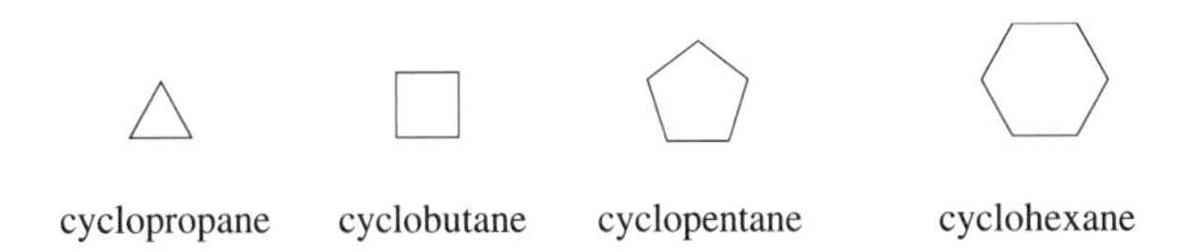

cyclopropane cyclobutane cyclopentane cyclohexane

2.2 Physical Properties

Alkanes have some general physical properties. The infrared spectrum for alkanes shows only two types of vibration. These are the carbon–carbon bond stretch and bending vibrations which are usually not useful as they are small peaks in the spectrum. The C–H (sp^3) bonds have distinct stretching peaks at around 2900 cm^{-1} and bending peaks at around 1400 cm^{-1}. Alkanes are characterised by an **absence of absorption**.

2.3 Industrial Preparations

Alkanes are readily available from the fractional distillation of crude oil. They are also obtained from natural gas. The natural alkanes obtained from oil can be modified by catalytic processes to produce other alkanes. Cracking, using Al_2O_3/SiO_2 at 450 °C, of long-chain alkanes gives short-chain alkanes and alkenes. Alkylation using concentrated sulfuric acid as a catalyst produces branched alkanes.

2.4 Laboratory Preparations

There are several laboratory preparations of alkanes; they all effectively rely on reduction of some functional group.

2.5 Preparation of Alkanes by Catalytic Hydrogenation of Alkenes or Alkynes

Catalytic hydrogenation of alkenes or alkynes is a versatile method which normally gives the product in quantitative (around 100%) yield.

$$R_2C{=}CR_2 \xrightarrow{H_2,\ \text{catalyst}} R_2CH{-}CHR_2$$

The catalysts most used are finely divided Pd or Pt or Adam's catalyst, PtO_2, which is reduced to platinum in the presence of hydrogen.

$$PtO_2 \xrightarrow{H_2} Pt$$

2.6 Preparation of Alkanes by Reduction of Alkyl Halides

It is possible to reduce an alkyl halide to an alkane. This can be achieved by several methods.

$$RX \longrightarrow RH$$

2.7 Preparation of Alkanes *via* the Grignard or the Alkyl Lithium Derivative

Alkyl halides can be converted to (a) Grignard reagents (RM_gX, where X is a halogen) or (b) lithium species (RLi) which react with water to produce an alkane.

$$\text{a)}\quad RX \xrightarrow{Mg} RMgX \xrightarrow{H_2O}$$

$$R{-}MgX + H{-}OH \longrightarrow RH + Mg(OH)X$$

$$\text{b)}\quad RX \xrightarrow{Li} RLi \xrightarrow{H_2O}$$

$$R{-}Li + H{-}OH \longrightarrow RH + LiOH$$

This is a useful method for introducing deuterium into a molecule.

$$RX \longrightarrow RD \qquad \textit{via}\ RMgX + D_2O$$

2.8 Preparation of Alkanes Using Hydride-reducing Agents and Alkyl Halides

There are many hydride-reducing agents such as $LiAlH_4$ or $LiBHEt_3$ (more soluble in an organic solvent). These will reduce most alkyl halides to the corresponding alkane.

$$\overset{+}{Li}H_3\overset{-}{Al}{-}H \quad R{-}X \xrightarrow{S_N2} RH$$

This works well for primary RX, moderately well for secondary RX and fails for tertiary RX.

2.9 Preparation of Alkanes by Reduction of Aldehydes and Ketones

See 10.23 and 10.24

2.10 Coupling of Alkyl Halides to Give Alkanes

Under certain conditions it is possible to couple two short alkyl metal (or aryl metal) derivatives to form a longer alkane. These alkyl or aryl metal derivatives can be formed from alkyl or aryl halides.

$$R^1X \xrightarrow{Li} R^1Li$$

$$2\,R^1Li \xrightarrow{CuI} R^1{-}Cu{-}R^1Li \xrightarrow{R^2{-}X} R^2{-}R^1$$

There is a high yield for any alkyl or aryl group R^1, given a primary R^2. It is a mediocre reaction for a secondary R^2 and fails for tertiary R^2. When R^2 is vinyl or phenyl the reaction works well, but not *via* a S_N2 mechanism as vinyl and phenyl cannot work through the S_N2 mechanism (see 8.15 to 8.18 for definitions of S_N2 mechanism).

2.11 Reactions of Alkanes

Alkanes are, relative to other functional groups, very inert. They are often used as solvents in reactions rather than reagents. In terms of reactions that give useful products, the main reaction of alkanes is halogenation.

2.12 Halogenation of Alkanes

All alkanes can be halogenated to some extent.

$$RH + X_2 \longrightarrow RX + HX$$

2.13 Chlorination of Methane

The chlorination of methane follows the reaction scheme below, where the chlorine atoms are introduced into the molecule in a stepwise fashion.

$$CH_4 + Cl_2 \xrightarrow{h\nu} CH_3Cl + HCl$$

$$CH_3Cl + Cl_2 \longrightarrow CH_2Cl_2 + HCl$$

$$CH_2Cl_2 + Cl_2 \longrightarrow CHCl_3 + HCl$$

$$CHCl_3 + Cl_2 \longrightarrow CCl_4 + HCl$$

The products from the reaction are a mixture of the four chloroalkanes. This is a useful reaction commercially, but of little importance in the laboratory.

2.14 Mechanism for the Chlorination of Methane

The mechanism through which the reaction proceeds is termed **a (free) radical chain** process.

1 $$Cl_2 \xrightarrow[\text{homolysis}]{h\nu} Cl\bullet + Cl\bullet \qquad \text{initiation}$$

2 $$Cl\bullet + CH_4 \longrightarrow CH_3\bullet + HCl$$

3 $$CH_3\bullet + Cl_2 \longrightarrow CH_3Cl + Cl\bullet$$

(steps 2 and 3: propagation)

There are generally three steps to the chain reaction mechanism: **initiation, propagation** and **termination**. The important reactant-consuming/product-forming steps are 2 and 3.

We can take a closer look at step 2:

$$Cl\bullet + H{-}CH_3 \longrightarrow CH_3\bullet + H{-}Cl$$

The C–H bond dissociation energy is 435 kJ mol^{-1}. Therefore

$\Delta H = +5\,kJ\,mol^{-1}$ (endothermic). The activation energy $\Delta G\# = 17$ KJmol^{-1}. An energy profile can be drawn.

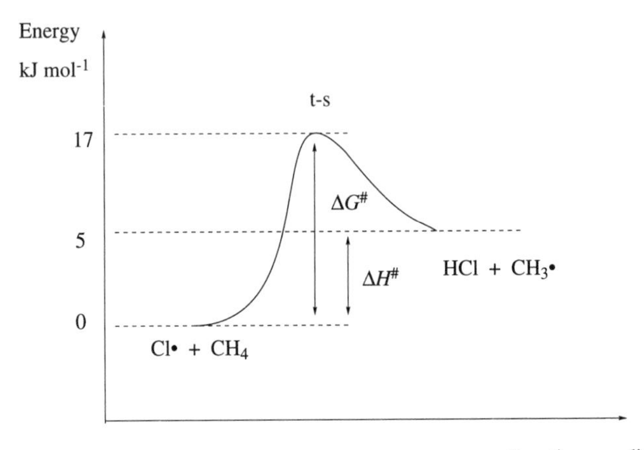

About one in every forty collisions provides enough energy to push the reactants over the energy barrier. The transition state, denoted by "#", can be thought of as drawn below.

We can also take a closer look at step 3:

$$CH_3\bullet \quad + \quad Cl-Cl \longrightarrow CH_3-Cl \quad + \quad Cl\bullet$$

$$Cl\bullet \quad H-CH_3 \longrightarrow \left[Cl\cdots H\cdots CH_3 \right]^{\#} \longrightarrow Cl-H \quad + \quad \bullet CH_3$$

In step 3 $\Delta G\#$ is very small and the reaction is very fast. Step 2 rather than 3 is the rate-limiting step.

2.15 Halogenation of Methane in General

Fluorine, F_2, is more reactive in the halogenation of methane than chlorine, Cl_2, and bromine, Br_2, is less reactive than chlorine. Iodine is so unreactive that under normal conditions there is no reaction.

The reason for this is that the overall rate is determined by the rate-limiting step (RLS).

$$X\bullet + CH_4 \longrightarrow CH_3\bullet + HX$$

C-H bond strength is 435 kJ mol^{-1}

X	F	Cl	Br	I
H-X bond strength kJ mol^{-1}	560	430	365	300
ΔH	-125	+5	+70	+135

decreasing strength of H-X bond, increasing ΔH and $\Delta G^{\#}$

The energy/reaction coordinate diagram below shows where the transition state occurs in the reaction, the higher the energy of the transition state, the later it occurs in the reaction.

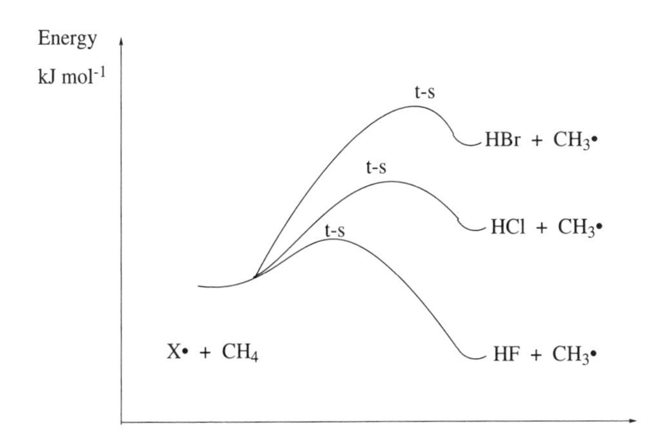

The transition states for the three halogenation reactions can be represented as something like those shown below.

$$\left[\overset{\bullet}{F}\text{-------H---}CH_3\right]^{\#} \qquad \left[\overset{\delta\bullet}{Cl}\text{-------H-------}\overset{\delta\bullet}{CH_3}\right]^{\#} \qquad \left[Br\text{--H---------}\overset{\bullet}{C}H_3\right]^{\#}$$

"reactant like" "in between" "product like"

2.16 Halogenation of Higher Alkanes

It is possible to halogenate higher alkanes, but it is often not as simple as the reactions shown below. Often multihalogenated compounds are formed.

$$CH_3CH_3 + X_2 \xrightarrow{h\nu} CH_3CH_2X$$

$$CH_3CH_2CH_3 + X_2 \xrightarrow{h\nu} CH_3CH_2CH_2X + CH_3CHXCH_3$$

X = Br: $CH_3CH_2CH_2X$ 3%; CH_3CHXCH_3 97%

The stability of carbon radicals, C•, has the order: tertiary are more stable than secondary, which in turn are more stable than primary, which are more stable than CH_3•. So:

1 $$Br_2 \xrightarrow{h\nu} 2Br\bullet$$

2 $$Br\bullet + CH_3CH_2CH_3 \xrightarrow{2a,\ -HBr} CH_3CH_2\overset{\bullet}{C}H_2 \xrightarrow{3a,\ Br_2} CH_3CH_2CH_2Br + Br\bullet$$

$$Br\bullet + CH_3CH_2CH_3 \xrightarrow{2b,\ -HBr} CH_3\overset{\bullet}{C}HCH_3 \xrightarrow{3b,\ Br_2} CH_3CHBrCH_3 + Br\bullet$$

$$\frac{\text{yield of } CH_3CH_2CH_2Br}{\text{yield of } CH_3CHBrCH_3} = \frac{\text{rate 2a}}{\text{rate 2b}} = \frac{3}{97}$$

The more stable radical is formed more quickly and therefore more of the 2-substituted product is formed.

If chlorine is used instead of bromine then 45% of 1-chloropropane is formed and 55% of 2-chloropropane. Bromine is less reactive than chlorine and therefore bromine is more selective than chlorine.

2.17 The Hammond Postulate

The structure of a transition state resembles the structure of the nearest stable species. Transition states for endothermic steps structurally resemble products, and transition states for exothermic steps structurally resemble starting materials.

2.18 Relevant Questions

42(a), 74(c), 151, plus parts of functional group interconversion questions.

3 ALKENES

3.1 General Features

Alkenes are **unsaturated** hydrocarbons containing a carbon–carbon double bond, C=C. Acyclic alkenes have the general formula C_nH_{2n}.

C_2H_4 $H_2C=CH_2$

C_3H_6 $CH_3CH=CH_2$

C_4H_8 has structural and geometrical isomers

Monocyclic alkenes have the general formula C_nH_{2n-2}.

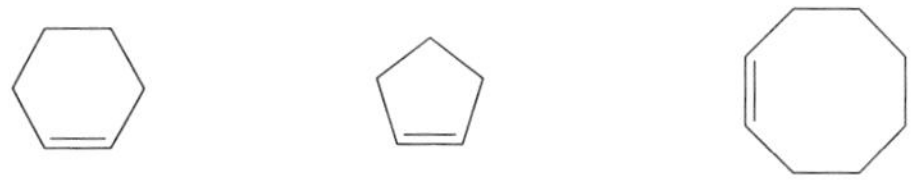

cyclohexene cyclopentene cyclooctene

The stability of alkenes increases as the number of alkyl groups attached to the double bond increases.

$H_2C=CH_2$ $RHC=CH_2$ $R_2C=CH_2$ $R_2C=CRH$ $R_2C=CR_2$

$RHC=CRH$

increasing stability due to hyperconjugation and steric hindrance

For a pair of *cis*/*trans* isomers, the *trans* isomer is usually the more stable. This is illustrated by the energy/stability diagram below, where ΔH is the enthalpy of hydrogenation.

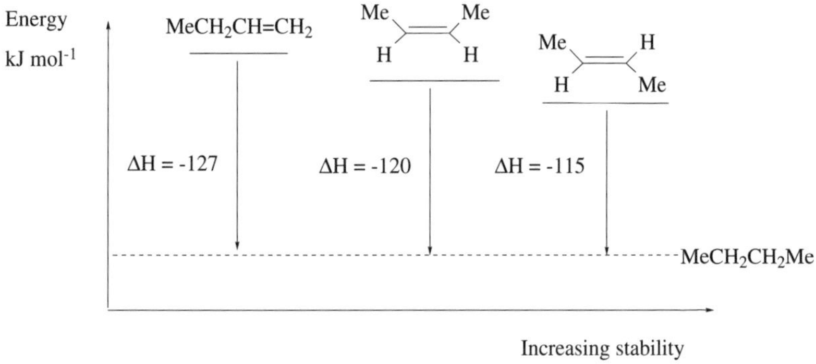

Alkenes have some general physical properties. The infrared spectra of alkenes have the vibrations shown by alkanes (sp^3 C–H vibrations) and sp^2 C–H vibrations. This gives rise to peaks at 3050 cm^{-1} and peaks around 650–1000 cm^{-1}. There is also a C=C bond vibration at 1600–1675 cm^{-1}; this peak has medium intensity.

3.2 Preparations

There are many preparations for alkenes. Some of the more common ones are given here.

3.3 Base-induced Elimination of Alkyl Halides

It is possible to form alkenes by base-induced elimination from alkyl halides and sulfonyl derivatives.

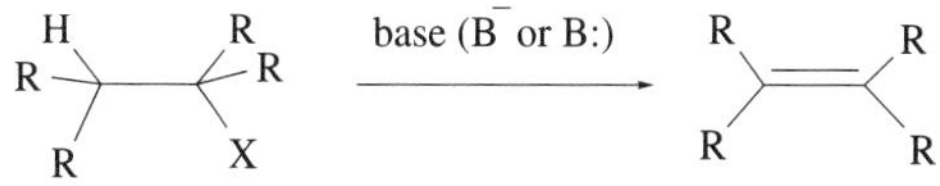

H R
R R base (B^- or B:) R R
R X R R

X = Cl, Br, I, OTs, OTf, OMs, OBs

base = conc. KOH in ethanol, RO^- in ROH or DBN (diazabicyclononene)

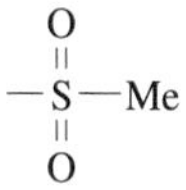

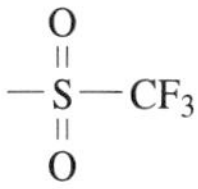

Ts $-SO_2-C_6H_4-$Me

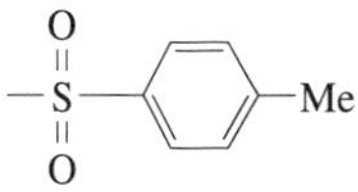

Ms $-SO_2-$Me

Tf $-SO_2-CF_3$

Bs $-SO_2-C_6H_4-$Br

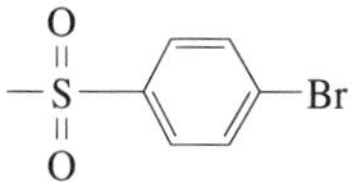

DBN

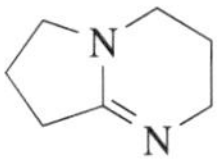

3.4 Mechanism for Base-induced Elimination of Alkyl Halides

(See Sykes 1984.)

There is a specific **stereochemical orientation** for the incoming base and the departing **leaving group** (or nucleofuge) in the so-called **E2** (elimination bimolecular) mechanism.

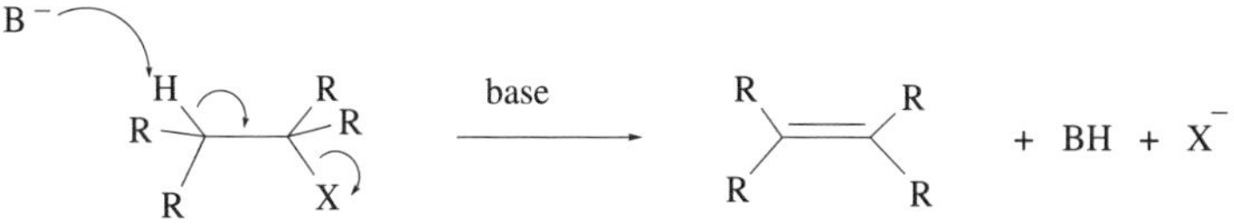

B^-
H R
R R base R R + BH + X^-
R X R R

this is a single-step process

The stereochemical orientation of the E2 mechanism requires an *anti*__periplanar__ (sometimes called *anti*coplanar) arrangement between the group that the base is removing (usually a proton) and the leaving group. This means that the leaving group and the proton are on opposite sides of the carbon–carbon bond (this is described by *anti*) and they must be in the same plane (described by *periplanar*). The result or process is called a **stereospecific** *anti* **elimination**.

B⁻ H R R R R β α X

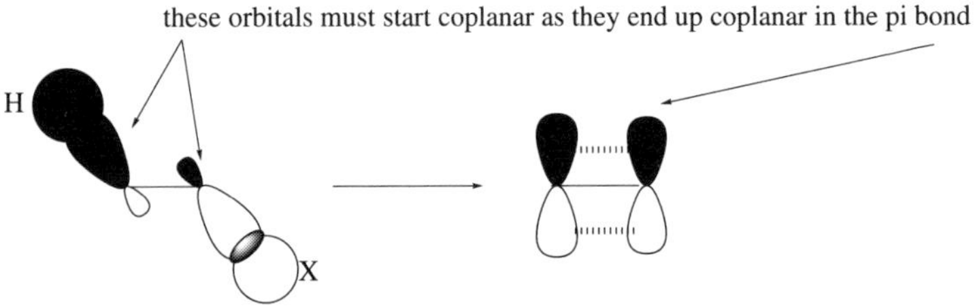

these orbitals must start coplanar as they end up coplanar in the pi bond

H X

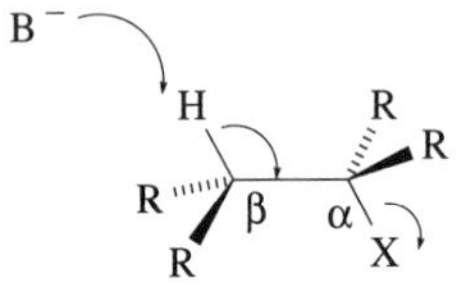

There must be an antiperiplanar (coplanar) configuration in order that the bonds being broken may reform to give the new pi bond. This entire process of stereospecific *anti* elimination is nicely demonstrated by the elimination of HBr in the following two examples.

H H Ph Me Ph Br — base → Me H Ph Ph only

H Ph H Me Ph Br — base → Me Ph Ph H only

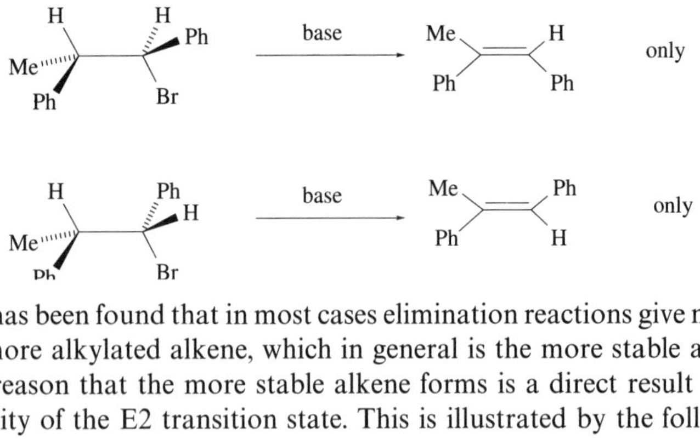

It has been found that in most cases elimination reactions give mainly the more alkylated alkene, which in general is the more stable alkene. The reason that the more stable alkene forms is a direct result of the stability of the E2 transition state. This is illustrated by the following two examples.

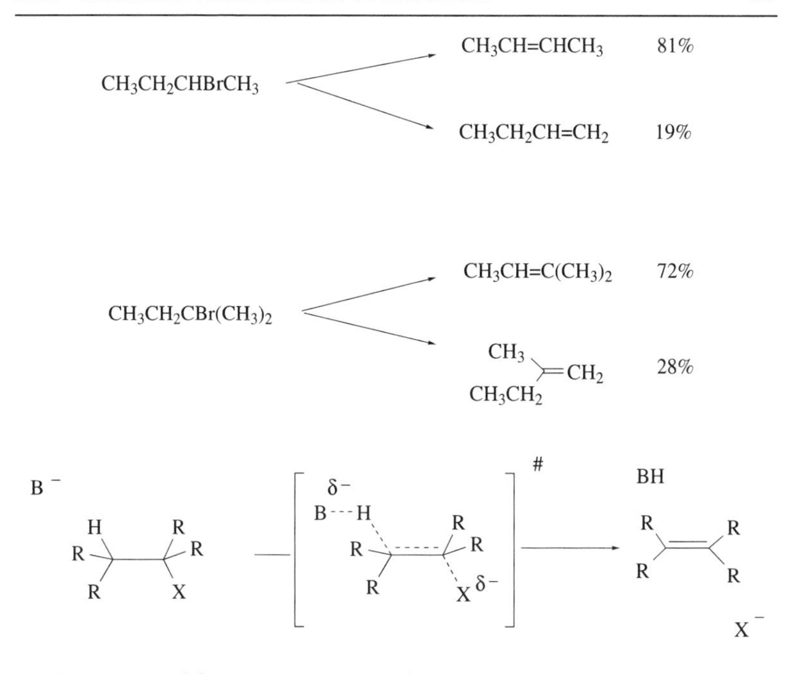

The E2 transition state (t-s) (transition states are normally drawn in brackets and marked with a #) has alkene character. The more stable transition state is the one that resembles the more stable alkene. Therefore the rate of reaction is faster for the more stable alkene.

3.5 Acid-catalysed Dehydration of Alcohols

The acid catalysed-dehydration of alcohols is another form of elimination. The molecule eliminated in this case is water.

$$R_2CH-CR_2(OH) \xrightarrow{\text{acid catalyst}} R_2C=CR_2 + H_2O$$

The ease with which the reaction proceeds is dependent on the nature of the carbon atom to which the hydroxyl group is attached. Tertiary alcohols eliminate better than secondary, which eliminate better than primary. The reactivities are illustrated by the conditions needed to give the alkene: for a primary alcohol you need concentrated sulfuric acid at

150 °C; for a secondary alcohol you need 50% sulfuric acid at 100 °C; for a tertiary alcohol you need 20% sulfuric acid at < 100 °C.

3.6 The E1 Elimination Mechanism for the Acid-catalysed Dehydration of Alcohols

(See Sykes 1984.)

The reaction only occurs under acid conditions not with base (HO^- is a **very poor** leaving group).

H R R R R OH H+ H R R R R OH2 + RLS

H R R R R + + H2O step 3 R R R R + H+

The reaction is in equilibrium and so the alkene must be distilled out as it is formed to drive the reaction to completion.

Rate of dehydration: step 2 is the rate-limiting step (RLS), which is the formation of the carbocation, a **reactive intermediate** in the reaction. The transition state for this step has carbocation character.

H R R δ+R R δ+OH2 #

the t-s has carbocation character

Carbocations have varying stabilities. In the simplest cases the stability is reflected by the series:

(most stable) tertiary > secondary > primary > Me (least stable)

For an alcohol, ROH, the stability of R^+ formed increases as we move from the primary to tertiary alcohol. The stability of the t-s increases accordingly, $\Delta G^{\#}$ decreases and the rate of dehydration of ROH increases.

Acid-catalysed dehydration gives mainly the more stable (more alkylated) alkene. The reason for this is step 3. The transition state for this step has alkene character. The more stable transition state (*i.e.* the more alkylated because it has alkene character) affords a faster reaction pathway. The less stable transition state affords a slower pathway.

$$MeCH_2CH(OH)CH_3 \longrightarrow MeCH{=}CHCH_3 + MeCH_2CH{=}CH_2$$

80% 20%

$R_2CH{-}C^+R_2 \longrightarrow [R_2C(H^{\delta+})\cdots CR_2^{\delta+}]^{\#} \longrightarrow R_2C{=}CR_2$

t-s has alkene character

ROH ⟶ R^+ — faster ⟶ [more stable t-s]# ⟶ more stable alkene, major product

ROH ⟶ R^+ — slower ⟶ [less stable t-s]# ⟶ less stable alkene

The correct terminology for a reaction where the formation of a double bond occurs after the nucleofuge has departed leaving a carbocation is **E1**. The E1 mechanism is a two-step process in which the rate-limiting step is the ionisation of the substrate (or leaving of the nucleofuge from the substrate). The first step of an E1 mechanism is the same as that of an S_N1 mechanism (see 8.12 and 8.14; also see March 1992 for a comprehensive discussion on the elimination mechanisms.)

3.7 Rearrangement of Carbocations

The carbocation intermediate sometimes rearranges. This is a 1,2-shift of the group R (alkyl, aryl or H). The equilibrium will lie to the side of the more stable carbocation.

secondary carbocation ⇌ tertiary carbocation

This is illustrated by the example below.

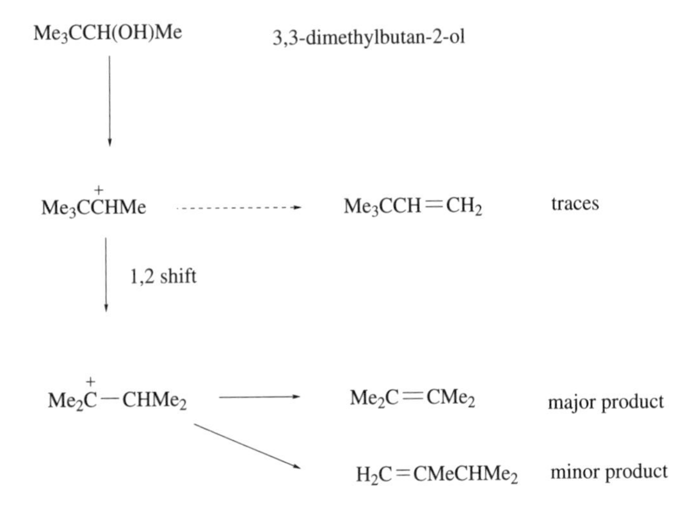

Dehydration of alcohols is not an ideal preparative reaction because forceful conditions are needed; also you do not always get the alkene you want because of the 1,2-shift.

3.8 Indirect (Two-step) Methods for Conversion of an Alcohol to an Alkene

An indirect method frequently used involves the conversion of the alcohol to a good leaving group, then the use of a base to effect the elimination.

$$\mathrm{ROH} \longrightarrow \mathrm{RX}$$

$$\mathrm{RX} \xrightarrow[\mathrm{E2}]{\text{base}} \text{alkene}$$

where X above is a halogen or sulfonyl group.

The conversion of an alcohol into an alkyl halide or a sulfonyl group (Ts, Ms, Tf or Bs), is covered in 5.22.

3.9 Direct (One-step) Methods for Conversion of an Alcohol to an Alkene

An effective direct method of converting an alcohol to an alkene involves the use of phosphorus oxychloride (phosphoryl chloride). Like most direct methods, a good leaving group is formed from the alcohol, which subsequently eliminates under the reaction conditions.
(See March 1992 for further examples.)

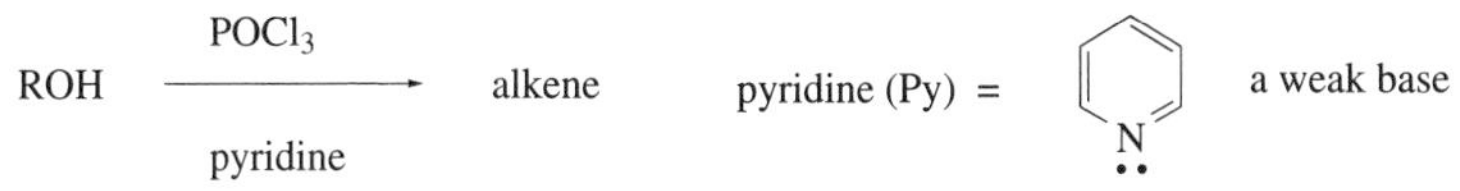

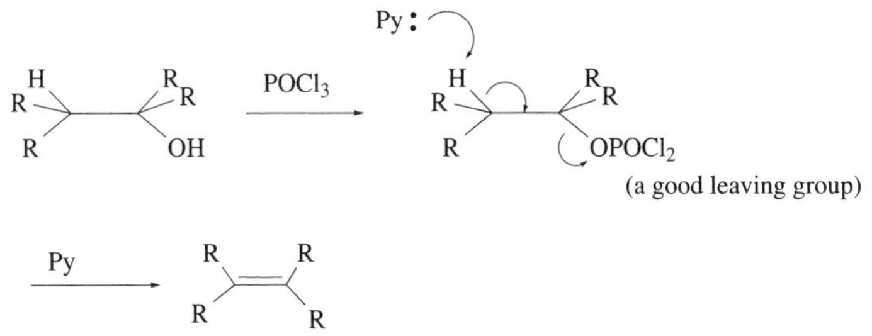

$$ROH \xrightarrow[\text{pyridine}]{POCl_3} \text{alkene}$$

pyridine (Py) = a weak base

(a good leaving group)

3.10 Alkenes by Thermal Cyclic Elimination of Esters

The thermal cyclic elimination of an ester goes *via* a **concerted** mechanism. There must be a proton on the α carbon.

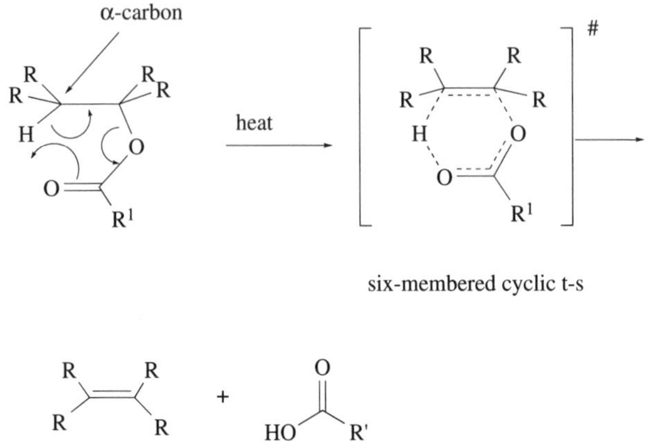

α-carbon

heat

six-membered cyclic t-s

The ester is vaporised and heated to 400–500 °C. No rearrangement of alkyl or aryl groups occurs, which implies there is no reactive intermediate. Stereospecific *syn* (same-side) elimination occurs.

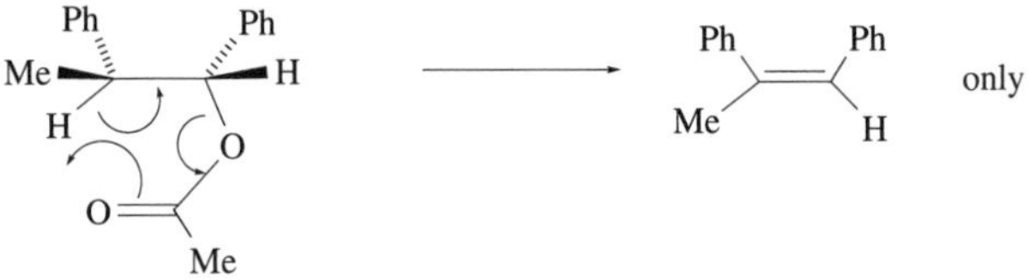

only

The middle carbon–carbon bond rotates so that the hydrogen and the ester group are in close proximity and have a planar arrangement. Hence the Ph and Me groups are on opposite sides in the alkene. This is necessary to produce the cyclic t-s.

3.11 The Wittig and Wadsworth–Emmons Reactions

The Wittig reaction is an important reaction for the synthesis of alkenes. Phosphoranes or phosphonium ylides react with aldehydes and ketones to produce alkenes.

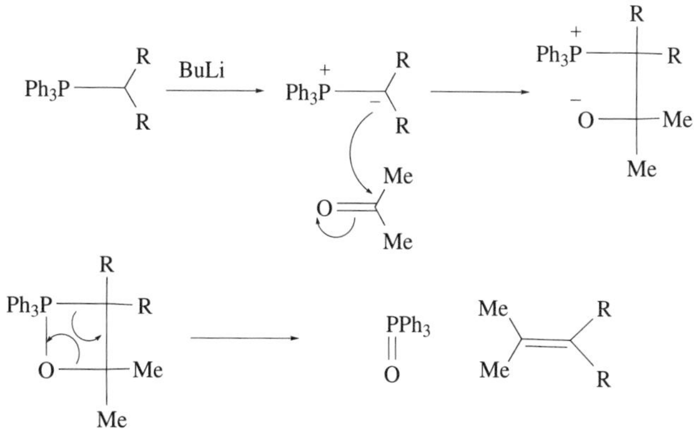

Formation of a strong P–O bond provides much of the driving force for the reaction. For the more stable ylides (R = aryl, COR, CO2R and CN) the ylide is less reactive and often a variant on the Wittig reaction, the Wadsworth–Emmons reaction, is used to ensure a more nucleophilic carbon. This modification of the Wittig reaction involves the use of a phosphonate anion which is a less stabilised and more reactive species.

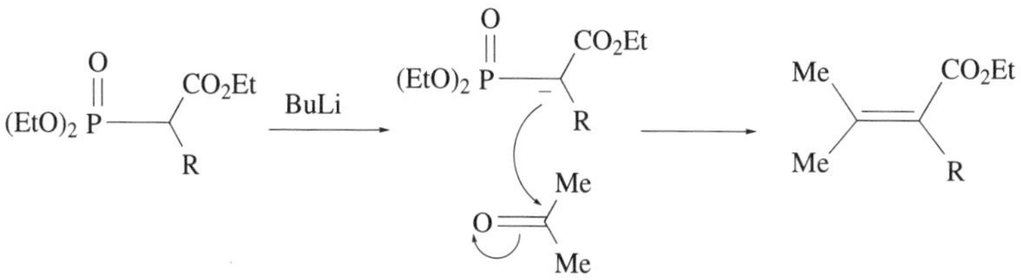

3.12 Alkenes by Partial Hydrogenation of Alkynes

Alkynes are molecules containing a carbon–carbon triple bond (see Chapter 4). It is possible to partially hydrogenate an alkyne and form an alkene. Using this method both *trans* and *cis* alkenes can be formed.

3.13 Alkenes by Catalytic Hydrogenation of Alkynes

Catalytic hydrogenation is carried out, as the name implies, using a catalyst and hydrogen. There are many hydrogenation catalysts.

$$\mathrm{RC{\equiv}CR} \xrightarrow[\text{catalyst}]{\mathrm{H_2}} \mathrm{RHC{=}CHR} \xrightarrow[\text{catalyst}]{\mathrm{H_2}} \mathrm{RH_2C{-}CH_2R}$$

However, a large number of catalysts cannot stop at the alkene stage, the alkyne being often reduced to the alkane. There are some catalysts that give the alkene: the most common one used is the Lindlar catalyst. This is a palladium on calcium carbonate catalyst doped with lead acetate. Quinoline is also added to the reaction mixture to facilitate the reaction.

The stereochemistry produced when using a catalyst and hydrogen is *syn* addition of hydrogen to the alkyne and so *cis* (with respect to the hydrogens) alkenes are produced. This is a direct effect of the mechanism by which a hydrogenation catalyst works.

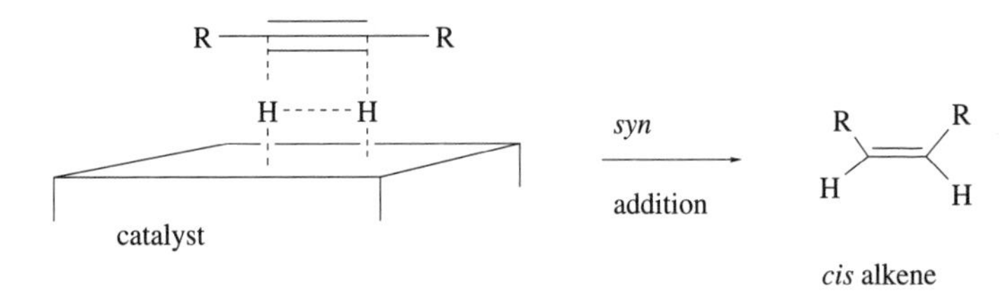

As can be seen above, the hydrogen is first absorbed on to the catalyst surface and is presented to the "same side" of the triple bond.

3.14 Alkenes from Alkynes Using Alkali Metals in Liquid Ammonia

Sodium or potassium in liquid ammonia provides a method of producing *trans* alkenes. There is *anti* addition of hydrogen due to the repulsion of the negative charges in the dianion to opposite sides of the molecule. When a source of protons is added (for example, aqueous NH_4Cl) the *trans* alkene is produced.

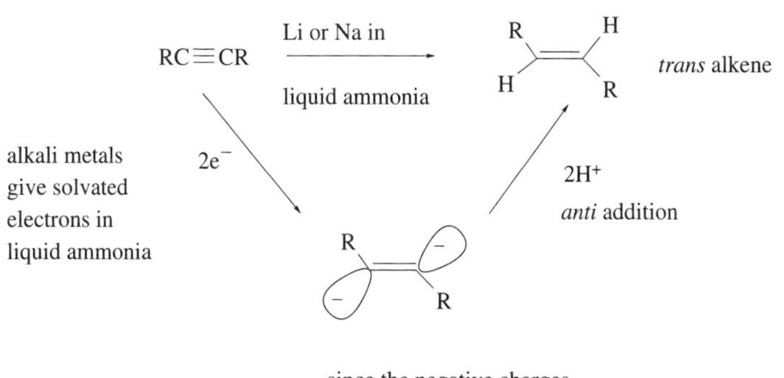

3.15 Reactions of Alkenes

There are a vast number of reactions concerning alkenes. A few of the more important ones are given in the following text.

3.16 Addition of Halogen Acids to Alkenes

Halogen acids (HCl, HBr, HI) can be added across a carbon–carbon double bond. The substituents on either side of the double bond determine the reactivity (effectively the rate of reaction) and the regioselectivity (which carbons the halogen and hydrogen end up bonded to) of a double bond towards the acid.

$$R_2C{=}CR_2 \xrightarrow{HX} R_2CH{-}CR_2X$$

HX = HCl, HBr, HI

Reactivity facts:

$PhCH{=}CH_2 > Me_2C{=}CH_2 > MeCH{=}CHMe > H_2C{=}CH_2$
$> H_2C{=}CHY$ ($Y = NO_2, CO_2H, CCl_3$)
most reactive least reactive

3.17 Regioselectivity of Addition of Halogen Acids to a Double Bond

The following three reactions show the regioselectivity of the addition of halogen acids across a double bond.

$$MeCH{=}CH_2 \xrightarrow{HX} MeCHX{-}CH_3 \quad (\text{no } MeCH_2CH_2X)$$

$$PhCH{=}CH_2 \longrightarrow PhCHX{-}CH_3$$

$$Me_2C{=}CH_2 \longrightarrow Me_2CX{-}CH_3$$

The regioselectivity in these reactions is summed up in Markovnikov's rule (generalisation): in an addition to an alkene the more negative part of the addendum goes to the carbon bearing the fewer hydrogens.

3.18 The Mechanism for Addition of a Halogen Acid to Alkenes

A mechanism that explains the regioselectivity has an electrophilic addition involving a carbocation intermediate.

R R RLS R R + R fast R R R
R R H R H X
δ+ δ- X⁻
H−X

The reactivity and regioselectivity depend on carbocation stability. The more stable the carbocation the faster it is formed.

3.19 Carbocation Stability in Simple Systems

The stability of carbocations has been shown to follow the order: allyl, benzyl > 3° alkyl > 2° alkyl > 1°alkyl > Me > vinyl, phenyl, where allyl or benzyl cations are the most stable and vinyl the least. The stability of carbocations follows the argument discussed in 3.34 and 3.39.

allyl

or

vinyl

benzyl

phenyl

For the three reactions in 3.17 the reactivity explanation is illustrated by the following:

$$PhHC{=}CH_2 \xrightarrow{H^+} Ph\overset{+}{C}H{-}CH_3 \text{ (benzyl)}$$

$$MeCH{=}CH_2 \longrightarrow Me\overset{+}{C}H{-}CH_3 \text{ (secondary)}$$

$$CH_2{=}CH_2 \longrightarrow \overset{+}{C}H_2{-}CH_3 \text{ (primary)}$$

carbocation stability decreases, therefore the reactivity of the alkene decreases

The regioselectivity explanation is found in the difference in relative rates for formation of the two possible carbocations.

$$MeCH{=}CH_2 \xrightarrow{a} Me\overset{+}{C}H{-}CH_3 \longrightarrow MeCHX{-}CH_3 \text{ major}$$

$$MeCH{=}CH_2 \xrightarrow{b} MeCH_2{-}\overset{+}{C}H_2 \longrightarrow MeCH_2{-}CH_2X \text{ negligible}$$

pathway (a) is *via* the more stable secondary cation and is therefore a faster pathway

pathway (b) is *via* the less stable primary cation and is subsequently slower

Addition goes *via* the more stable carbocation intermediate.

3.20 Anomalous Addition of HBr to an Alkene when a Radical Source is Present

Normal electrophilic addition is:

$$\mathrm{MeCH{=}CH_2} \xrightarrow{\mathrm{HBr}} \mathrm{MeCHBrCH_3}$$

However, radical addition is possible.

$$\mathrm{MeCH{=}CH_2} \xrightarrow[\text{trace of peroxide}]{\mathrm{HBr}} \mathrm{MeCH_2CH_2Br} \qquad \text{very fast reaction}$$

This occurs much faster.

3.21 Mechanism for Addition of HBr to an Alkene when a Radical Source is Present

The radical mechanism is, when ROOH is the peroxide (a radical source),

$$\left.\begin{array}{l} \mathrm{RO{-}OH} \longrightarrow \mathrm{RO{\bullet}} + {\bullet}\mathrm{OH} \\ \mathrm{RO{\bullet}} + \mathrm{HBr} \longrightarrow \mathrm{ROH} + \mathrm{Br{\bullet}} \end{array}\right\} \text{initiation}$$

$$\left.\begin{array}{ll} \text{1)} & \mathrm{Br{\bullet}} + \mathrm{MeCH{=}CH_2} \longrightarrow \mathrm{Me\dot{C}HCH_2Br} \\ \text{2)} & \mathrm{Me\dot{C}HCH_2Br} + \mathrm{HBr} \longrightarrow \mathrm{MeCH_2CH_2Br} + \mathrm{Br{\bullet}} \end{array}\right\} \text{radical chain}$$

Some facts about the reaction are:

1. Regioselectivity; the alkene is attacked initially by Br• to give the more stable radical intermediate. (The order of stability of C• is the same as that of C^+.)
2. The HBr radical mechanism is fast: it dominates the electrophilic mechanism if a peroxide (or other radical source) is present.
3. For HCl and HI the radical mechanism is slow. The electrophilic mechanism always dominates.

For HI step 1 is slow because I• is unreactive.
For HCl step 2 is slow because the H–Cl bond is strong.

3.22 The Use of Water and Sulfuric Acid in the Hydration of Alkenes

$$RCH{=}CH_2 \xrightarrow{H_2SO_4} R\overset{+}{C}H{-}CH_3 \;\; {}^{-}OSO_2OH \longrightarrow RHC(OSO_2OH){-}CH_3$$

$$RCH{=}CH_2 \longrightarrow R\overset{+}{C}H{-}CH_3 \; (OH_2) \overset{H^+}{\rightleftharpoons} RHC(OH){-}CH_3$$

$$RHC(OSO_2OH){-}CH_3 \longrightarrow RHC(OH){-}CH_3$$

In this reaction there is Markovnikov orientation. The carbocation can often rearrange and the conditions of the reaction are vigorous, so usually complex molecules do not survive.

3.23 Addition of Halogens to Alkenes

Halogens can be added across a double bond.

$$R_2C{=}CR_2 \xrightarrow[\text{solvent}]{X_2} R_2CX{-}CXR_2$$

$X_2 = Cl_2, Br_2$, (I_2 for some alkenes)

solvent = ethanoic acid or tetrachloromethane

Some facts about the reaction are:

1. Alkene reactivity order is the same as addition of HX.
2. Addition of another type of halide ion results in mixed dihaloalkanes being formed.

$$R_2C{=}CR_2 \xrightarrow[\text{solvent, LiCl}]{Br_2} R_2CBr{-}CBrR_2 + R_2CCl{-}CBrR_2$$

3. We get stereospecific *anti* addition.

Me, H, Me, H $\xrightarrow{Br_2}$ Me, H, Br, Br, Me, H + H, Br, Me, Br, H, Me

Me, H, H, Me $\xrightarrow{Br_2}$ Me, H, Br, H, Me, Br

trans *meso*

3.24 Mechanism for Addition of Halogens to Alkenes

There is electrophilic addition, with a **halonium ion** intermediate.

H, H, Me, Me; Br δ+ — Br δ- ⟶ H, H, Me, Me, Br+; Br⁻ ⟶ H, Me, Br, Br, H, Me

the Br can now attack from behind, S_N2 fashion with inversion of configuration

H, H, Me, Me, Br+ a bromonium ion intermediate

There are a series of related additions involving

$\overset{\delta+}{Br}-\overset{\delta-}{Cl}$ $\overset{\delta-}{I}-\overset{\delta+}{Cl}$ $\overset{\delta-}{HO}-\overset{\delta+}{Cl}$ $\overset{\delta+}{Br}-\overset{\delta-}{OH}$

Where there is a choice of attack the nucleophile will attack at the carbon with more positive charge on it.

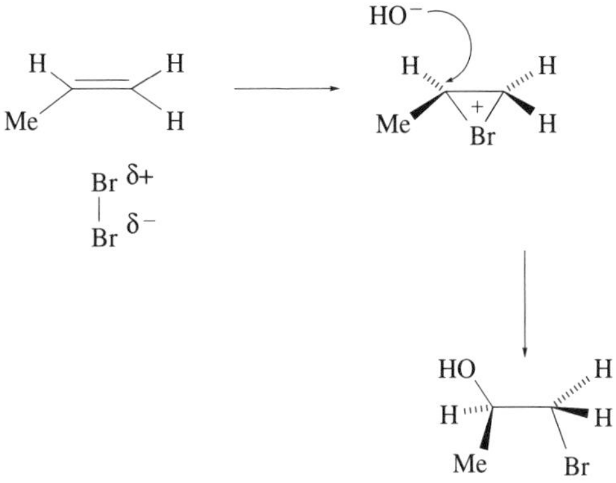

the $\overline{HO}$ attacks the carbon with the methyl attached because it has more positive charge on it

3.25 Oxidative Addition (Dihydroxylation) of Alkenes

There are two possible outcomes from a dihydroxylation of a double bond.

3.26 *Anti* Dihydroxylation of Alkenes

In *anti* dihydroxylation the hydroxyls can be thought of as having ended up on "different sides" of the molecule. The reaction proceeds through the stable (isolable) epoxide. The peroxy acid used is often *m*-chloroperbenzoic acid (MCPBA).

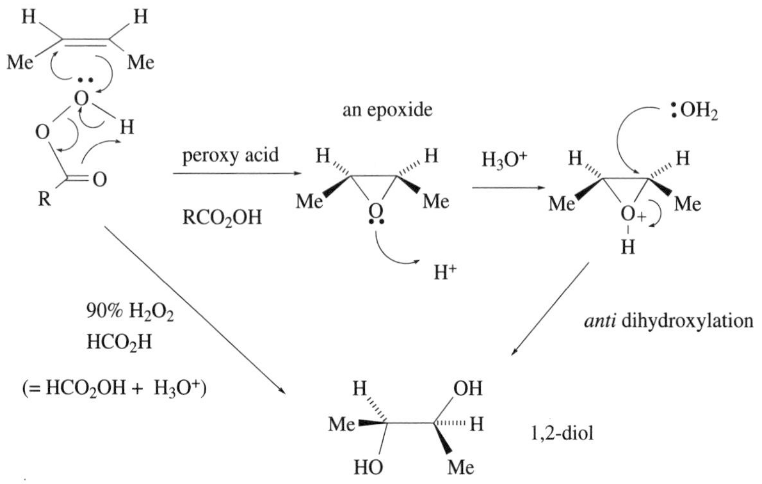

an epoxide

peroxy acid

RCO_2OH

H_3O^+

H^+

90% H_2O_2
HCO_2H

(= HCO_2OH + H_3O^+)

anti dihydroxylation

1,2-diol

3.27 *Syn* Dihydroxylation of Alkenes

In *syn* dihydroxylation the hydroxyls can be thought of as ending up on the "same side" of the molecule.

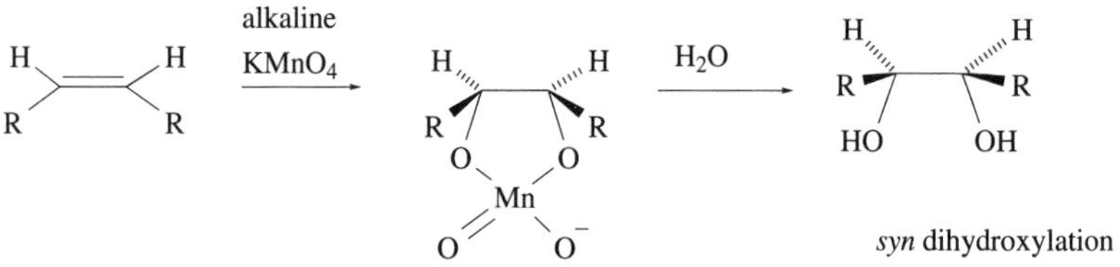

alkaline $KMnO_4$

H_2O

syn dihydroxylation

3.28 Oxidative Degradation: Indirect Cleavage of Alkenes

An alkene can first be transformed into a diol, as above, and then cleavage of the carbon–carbon bond can take place using a periodate.

dihydroxylation

HIO_4

aldehydes and/or ketones depending on R

3.29 Oxidative Degradation: Direct Cleavage of Alkenes

(See March 1992 for more details on ozonide intermediates.)

Alternatively, a double bond can be cleaved directly by the use of ozone. The reaction proceeds through several intermediates, one of which is an ozonide.

O_3 in O_2

ozonide

aq. $MeCO_2H$

zinc

$+ H_2O_2$

aldehydes and/or ketones depending on R

the Zn destroys the H_2O_2 produced which attacks aldehydes

3.30 Indirect Addition of Water: Markovnikov

Water can be added across a double bond to give **Markovnikov** addition. Mercury diacetate first adds across the double bond in an oxymercuration step. Water then displaces the mercury acetate group. The mechanism involves the ring opening of a mercurinium ion by S_N2

attack of water; this results in the *anti* (or *trans*) hydroxy-substituted organomercury, which is then reduced to give the alcohol.

$Hg(OAc)_2$ / H_2O → reduction → + Hg

mercurinium ion

two-step oxymercuration

Markovnikov addition of H_2O, no rearrangements

3.31 Indirect Addition of Water: anti-Markovnikov

Water can also be added across a double bond to give **anti-Markovnikov** addition. A borane adds across the double bond such that the boron becomes bonded to the **less substituted** carbon. The stereochemistry of the reaction requires the hydrogen and boron components to add to the "same side" of the double bond in a *syn* (or *cis*) fashion The boron is then oxidatively cleaved, leaving the hydroxyl.

BH_3 / ether, hydroboration → not isolated → H_2O_2 / ^-OH

anti-Markovnikov addition of H_2O

BH_3 / ether → H_2O_2 / ^-OH

from "BH_3"

3.32 Allylic Halogenation

The double bond in an alkene can aid reactions at other positions in the molecule. This is the case with allylic halogenation. The allylic carbon is halogenated using *N*-bromosuccinimide (NBS). This is a radical reaction initiated by trace amounts of Br_2 in NBS. The function of the NBS is to provide Br_2 and to use up HBr liberated in the reaction.

$$CH_2{=}CH_2CH_3 \xrightarrow[\text{CCl}_4 \text{ peroxide or light}]{\text{NBS}} CH_2{=}CH_2CH_2Br$$

NBS (*N*-bromosuccinimide) = [structure: five-membered ring with two C=O groups flanking N–Br]

[cyclohexene] $\xrightarrow[\text{CCl}_4 \text{ peroxide or light}]{\text{NBS}}$ [3-bromocyclohexene] not [1-bromocyclohexene] or [4-bromocyclohexene]

Here we have allylic substitution of Br for H. The H replaced is an allylic one (*i.e.* on a carbon next to a double bond).

3.33 Mechanism of Allylic Halogenation

A simple radical mechanism is involved. The double bond stabilises the radical at the allylic position.

$$CH_2{=}CH_2CH_3 \xrightarrow[\text{H}\bullet]{\text{removal of}} CH_2{=}CH_2\dot{C}H_2 \xrightarrow[\text{of Br}\bullet]{\text{addition}} CH_2{=}CH_2CH_2Br$$

3.34 Origins of Stabilisation of Allylic Radicals

Why are allylic radicals stabilised? There are two approaches to explain the stabilisation. They are just different models.

Orbital approach: This is probably the more accurate description of what is really happening to stabilise the allylic radical, an understanding of molecular orbital theory helps here (see Gray 1973 and Cartmell and Fowles 1979). Effectively the two p orbitals which initially form the pi bond of the alkene combine with the p orbital initially holding the single electron. A new orbital system is formed, the result of which allows a lowering of the overall energy of the molecule compared to a radical which cannot overlap with a pi system. The radical A, below, can be formed and is adjacent to a pi system.

A

We could view this is as the single electron delocalised over all three carbons in this new orbital system.

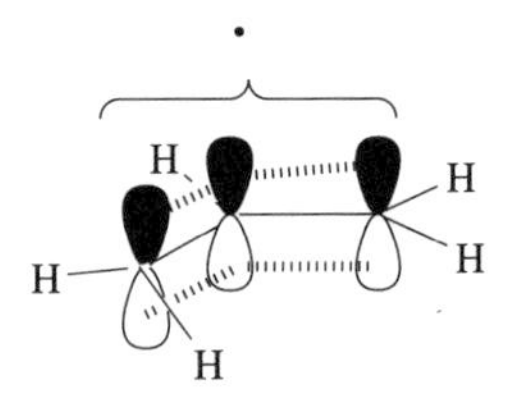

single electron delocalised over the three orbitals

This is more stable than the simple $CH_3CH_2CH_2\bullet$ radical, which cannot form a new orbital system through overlap with a pi bond.

Valence bond description approach: This requires an acceptance that the more resonance structures (canonicals) a system has, the more stable it is.

On forming the radical it can be localised on either A or B. A and B are resonance canonicals; note that the radical is never on the central carbon.

$$CH_2{=}CHCH_3 \longrightarrow \left[CH_2{=}CH\dot{C}H_2 \longleftrightarrow \dot{C}H_2CH{=}CH_2 \right]$$

A B

The representations A and B together are equivalent to:

$$\overbrace{H_2C \overset{\bullet}{=\!=} C =\!= CH_2}$$

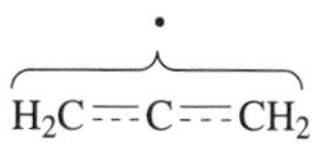

Care must be taken with this description as the pictorial representations of the resonance canonicals could imply that the species A and B exist as discrete entities; they do not. The allyl radical is a **resonance hybrid** of the contributing structures A and B.

The radical $CH_3CH_2CH_2\bullet$ has no resonance canonicals and is therefore less stable than the allyl radical.

A similar circumstance exists for allyl cations; the allyl cation C^+ is resonance-stabilised.

$$CH_2{=}CHCH_3 \longrightarrow \left[CH_2{=}CH\overset{+}{C}H_2 \longleftrightarrow \overset{+}{C}H_2CH{=}CH_2 \right]$$

The representations C and D together are equivalent to:

$$\overbrace{H_2C =\!= \underset{H}{C} =\!= CH_2}^{+}$$

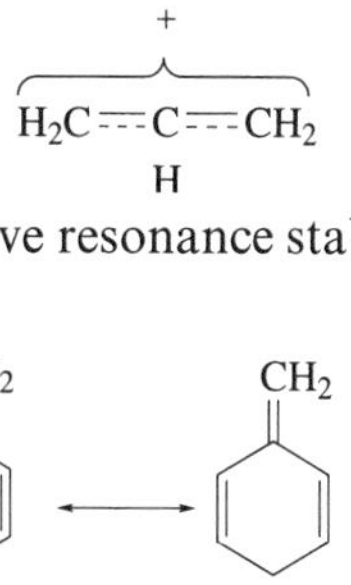

Benzyl C• and C^+ also have resonance stabilisation:

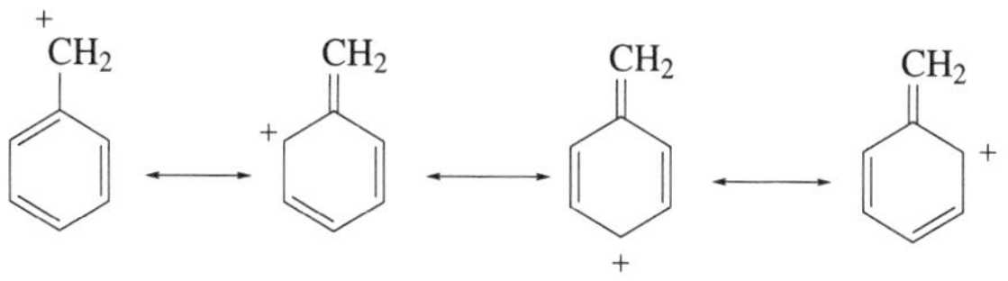

$$C_6H_5\overset{+}{C}H_2 \longleftrightarrow \longleftrightarrow \longleftrightarrow$$

Again, in all three of these latter systems, there is also a molecular orbital description.

Vinyl and phenyl C• and C^+ are not resonance-stabilised and so are relatively hard to form compared to allyl or benzyl systems. In terms of molecular orbital theory, no new orbital system is obtained by removing one or two electrons from an existing orbital system, there is only destabilisation of the existing orbitals.

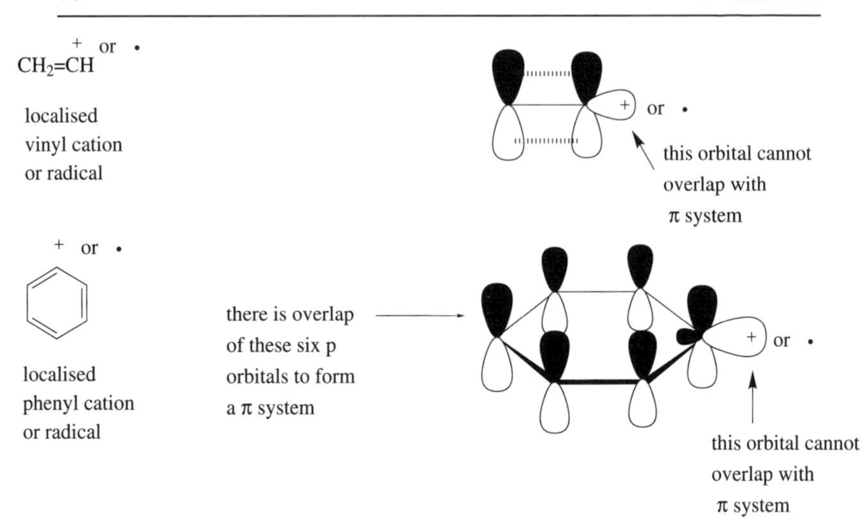

In summary: allyl and benzyl radicals and cations are more stable than alkyl, which are more stable than vinyl and phenyl.

3.35 Conjugation of Double Bonds

Conjugated double bonds occur when a compound has alternating double and single bonds.

3.36 Electronic Structure of Conjugated Double Bonds

For 1,3 butadiene there is pi-type overlap of the pi orbitals across C2 and C3. Another way of looking at this is to consider the original four molecular p orbitals all combining to form four new orbitals. These new orbitals are four pi molecular orbitals: two bonding and two antibonding (see Gray 1973 and Cartmell and Fowles 1979).

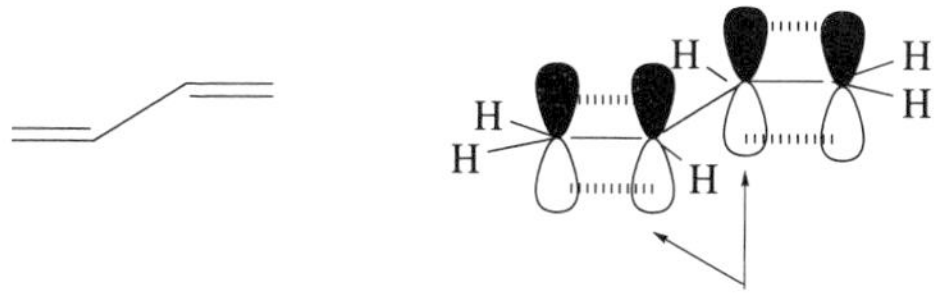

H H H H H H

partial overlap of these two pi systems a C2 and C3

As a consequence of this overlap, or formation of a new orbital system, conjugated dienes are more stable than non-conjugated ones. The approximate heats of hydrogenation show this: less energy is evolved by the conjugated system than by the non-conjugated, therefore it must contain less energy, *i.e.* it is more stable.

	ΔH hydrogenation (kJ mol^{-1})
$CH_2{=}CHR$	ΔH = -127
$CH_2{=}CHCH_2CH{=}CH_2$	ΔH = -254
$CH_2{=}CHCH{=}CH_2$	ΔH = - 238

Conjugated systems are planar and eclipsed about C2 and C3 (see 1.8). This is a requirement in order to obtain overlap of the pi orbitals (see Williams and Fleming 1995 for details of UV spectroscopy of 1,3-butadiene).

3.37 1,2 and 1,4 Addition to Dienes

When a diene is exposed to a halogen or a halogen acid (or another reagent which can add across a double bond) either 1,4 or 1,2 addition can occur to give two products.

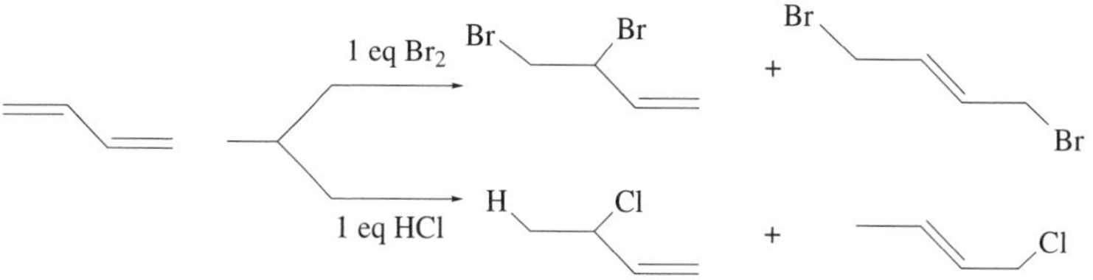

1 eq Br_2 Br Br + Br Br

1 eq HCl H Cl + Cl

When HCl is added across the double bond, the **homoallyl** (homoallyl means associated with the second carbon from the double bond) carbocation is not formed, because it is unstable relative to both the alternative allyl cations that can be formed. The allyl cations proceed further to form the products:

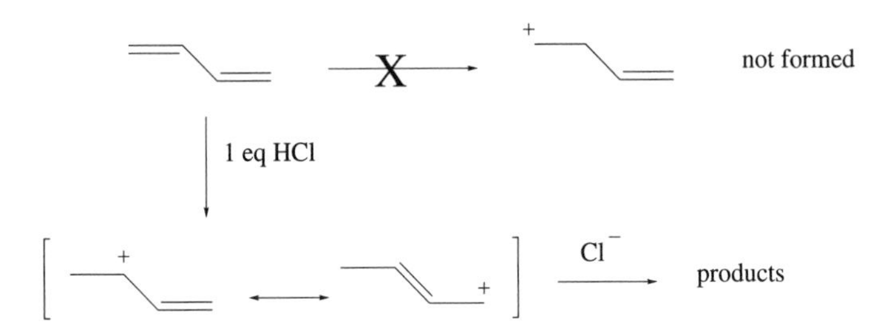

3.38 Kinetic and Thermodynamic Control of Reactions

Reactions are either kinetically controlled or thermodynamically controlled. Kinetic control operates if there is no equilibrium in the reaction (*i.e.* the reaction is not reversible). In this case the product ratio A : B is equal to:

the rate of formation of A : the rate of formation of B

or

$K_A : K_B$.

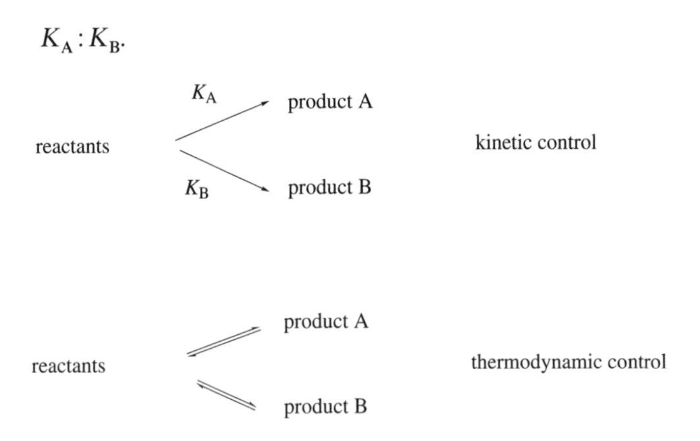

If however, (1) the products can interconvert (*i.e.* are formed reversibly) and (2) the system is allowed to reach equilibrium, then the ratio of A : B is determined by the stability of A compared to B and is then under thermodynamic control.

For a 1,3-diene reacting with HCl to give the 1,2 or 1,4 addition product the reaction can be under either kinetic or thermodynamic control depending on the temperature at which the reaction is carried out.

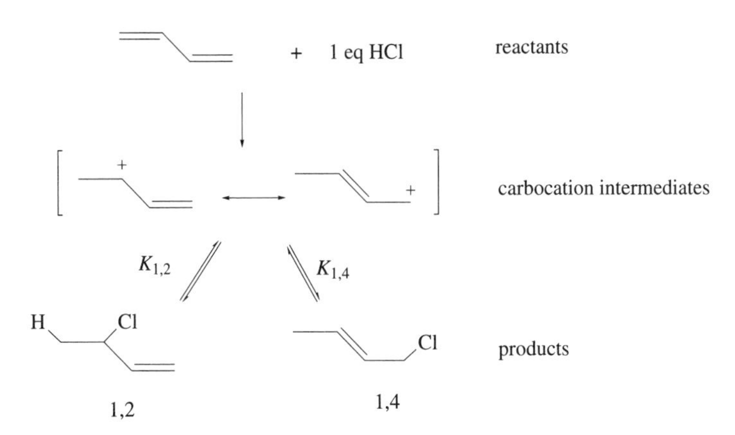

The ratio of products at two different temperatures are shown below. At the lower temperature the proportions in which they are isolated show the proportions in which they were initially formed. As the temperature is raised, the proportions in which the products are initially formed may remain the same, but there is faster conversion of the initially formed products into the equilibrium mixture.

	1,2 :	1,4	
−40 °C	76	24	kinetic control ($K_{1,2} > K_{1,4}$)
+80 °C	25	75	thermodynamic control (1,4 more stable than 1,2)

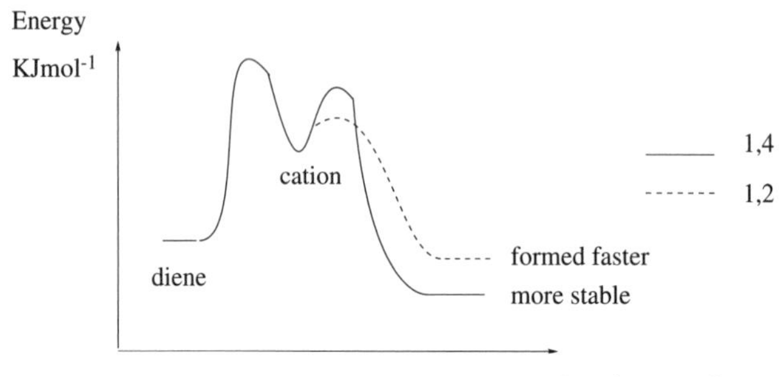

The reason for this is that at low temperatures the products cannot revert back to the carbocation, C^+, and Cl^-, so equilibrium is not established.

3.39 Hyperconjugation

The stability of carbon radicals, C•, and carbocations, C^+, has the order; 3° alkyl > 2° alkyl > 1°alkyl > Me. This is partially due to hyperconjugation, where sigma bonds have an overlap with the orbital containing the lone electron or positive charge.

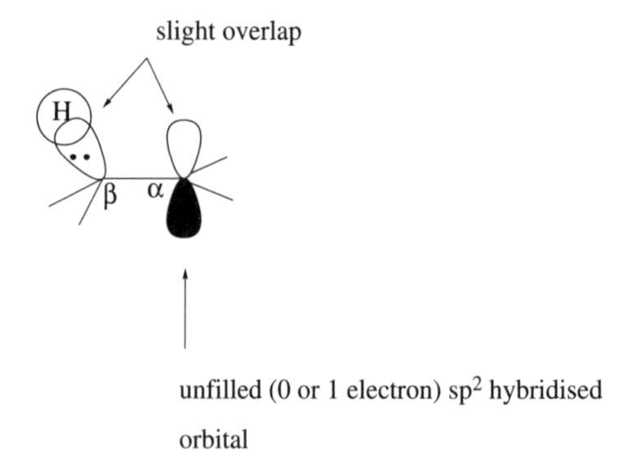

There is a slight overlap of electrons in a C_β–H bond with the unfilled p orbital on C_α. There is therefore a slight delocalisation of the unpaired electron or positive charge onto C_α, and hence there is a slight stabilisation of C• or C^+. The delocalisation and stabilisation is proportional to the number of C_β–H bonds.

$(CH_3)_3C^{\bullet}$ — 9 C_{β}-H; $(CH_3)_2CH^{\bullet}$ — 6 C_{β}-H; $CH_3CH_2^{\bullet}$ — 3 C_{β}-H; $^{\bullet}CH_3$ — 0 C_{β}-H

decreasing stabilisation →

With C^+ there is also a small positive inductive effect (+I) of alkyls relative to H. This is the second factor contributing to stability.

$(CH_3)_3C^+$; $(CH_3)_2CH^+$; $CH_3CH_2^+$; $^+CH_3$

decreasing stabilisation by hyperconjugation and the +I effect →

3.40 The Diels–Alder Reaction

The Diels–Alder reaction is a reaction between almost any conjugated diene and any double bond. It is a widespread and fundamental reaction in organic chemistry.

conjugated diene + alkene (dienophile) (R) → six-membered cyclic adduct (R)

The dienophile (or double bond) reacts faster if it is conjugated to C=O or CN and a Lewis acid (*e.g.* $AlCl_3$) is used as a catalyst.

$CH_2{=}CH{-}CHO$, $CH_2{=}CH{-}CO_2Et$, maleic anhydride (O, O, O) are good dienophiles

3.41 Mechanism of the Diels–Alder Reaction

The reaction is thought to be concerted (*i.e.* there are no intermediates).

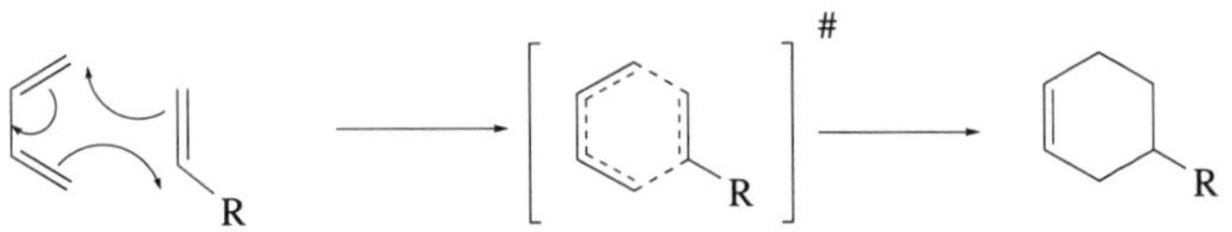

six-membered t-s

The most reactive type of conjugated dienes are cyclic ones, where the diene is held in the correct conformation for reaction to take place. Those dienes held in the wrong conformation do not react.

does not react, held in wrong conformation

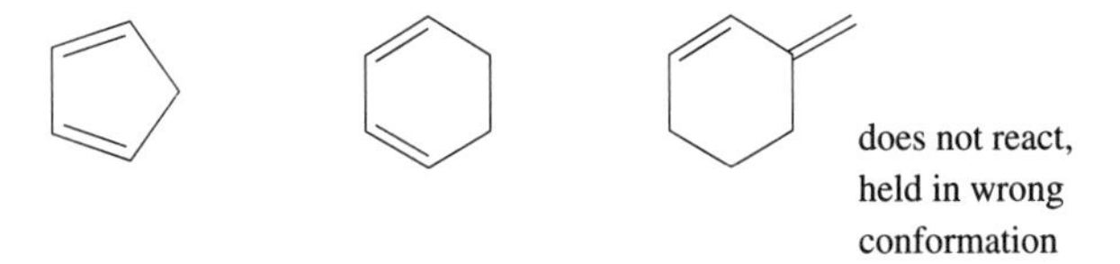

3.42 Relevant Questions

3, 31, 33, 38, 40, 44, 54(i), 64, 70(a), 90, 94, 115(b), 135, 137(a), 138(b), 143(a), 157, 158, 170, plus parts of functional group interconversion questions.

4 ALKYNES

Alkynes are unsaturated hydrocarbons containing a carbon–carbon **triple** bond (acetylenic bond).

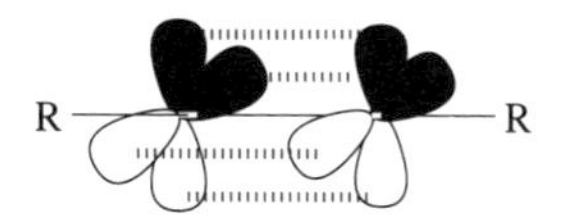

Acyclic alkynes have the general formular C_nH_{2n-2}. The simplest is C_2H_2 or ethyne.

$H-\!\equiv\!-H$	ethyne
$H_3C-\!\equiv\!-H$	propyne
$CH_3CH_2-\!\equiv\!-H$	but-1-yne, a terminal alkyne
$H_3C-\!\equiv\!-CH_3$	but-2-yne, non-terminal alkyne

4.1 Physical Properties of Alkynes

The infrared spectrum shows a terminal C–H stretch at 3300 cm^{-1}. There is also a weak C-C triple bond stretch at 2260–2100 cm^{-1}.

4.2 Preparation of Alkynes from Alkenes

By dihalogenation of a double bond, followed by elimination of two moles of hydrogen halide it is possible to form a triple bond. Usually bromine is used as the halogen. It is easier to dihalogenate a *cis* alkene than a *trans*.

R(H)C=C(R)H → (Br_2) → RCH(Br)–CH(Br)R → (base, loss of 2HBr) → R–≡–R

cis or *trans*

the base used is usually KOH/EtOH or $NaNH_2$/liquid NH_3

4.3 Preparation of Alkynes from Lower Alkynes

Lower **terminal** alkynes have an acidic proton, which can be removed using a strong base. The resulting anion is a good nucleophile and will displace certain groups, including halogens. This is a good method for forming unsymmetrical alkynes.

H–≡–H → ($NaNH_2$) → H–≡$^-$ Na^+ → (R^1X) → H–≡–R^1

(X = halogen, S_N2 reaction)

→ ($NaNH_2$, R^2X) → R^2–≡–R^1

The reaction is good when R^1X and/or R^2X are primary, it may work if they are secondary and is useless when they are tertiary.

4.4 Reactions of Alkynes

Many of the reactions of alkynes are akin to those of alkenes, such as elctrophilic addition, oxidative degradation and the Diels–Alder reaction. However, alkynes have a set of reactions which alkenes do not have because of their acidity.

4.5 Hydrogenation of Alkynes

Alkynes can be hydrogenated to give either alkenes or alkanes. On the laboratory scale both are important synthetic reactions. See 3.13 for further details.

R—≡—R $\xrightarrow{H_2,\ cat.}$ (H)(R)C=C(H)(R) $\xrightarrow{H_2,\ cat.}$ RHHC—CHHR

4.6 Electrophilic Addition to Alkynes

It is possible to add one or two moles of bromine across a triple bond.

R—≡—R $\xrightarrow[1\ eq]{Br_2}$ (R)(Br)C=C(Br)(R) $\xrightarrow[adds\ slower]{Br_2}$ R(Br)(Br)C—C(Br)(Br)R

R—≡—H $\xrightarrow{HCl}$ (R)C⁺=C(H)(H) Cl⁻ ⟶ (R)(Cl)C=C(H)(H)

$\xrightarrow{HI}$ (Cl)(R)C⁺—CH₃ I⁻ ⟶ (Cl)(R)(I)C—CH₃

4.7 Addition of Water to Alkynes

Addition of water to an alkyne results in either an aldehyde or a ketone being formed, depending on whether the addition is "Markovnikov" or "anti-Markovnikov" in nature.

4.8 Markovnikov Addition of Water to Alkynes

R—≡—H $\xrightarrow[HgSO_4]{H_2SO_4\ (aq)}$ (R)(HO)C=C(H)(H) ⇌ R—C(=O)—CH_3

R—≡—H $\xrightarrow{Hg^{2+}}$ H_2O: attacks mercurinium ion (R, H, Hg^{2+}) ⟶ (HO)(R)C=C(H)(Hg^+) $\xrightarrow{H_3O^+}$ (R)(HO)C=C(H)(H)

4.9 Anti-Markovnikov Addition of Water to Alkynes

R—≡—H →(BH_3) R(H)C=C(H)(BH_2) →(HO^- / H_2O_2) R(H)C=C(H)(OH) ⇌ RCH_2CHO

R—≡—H ↓($R^1{}_2BH$) R(H)C=C(H)B(R^1)(R^1) →(HO^-/H_2O_2) R(H)C=C(H)(OH)

groups with large steric hindrance

It is difficult to stop at the mono-addition of BH_3. It is likely that $RCH_2CH(BH_2)_2$ is formed. If BH_3 is replaced by $R^1{}_2BH$ where R^1 is bulky, then only mono-addition is observed.

4.10 Oxidative Degradation of Alkynes

Ozonolysis of alkynes produces carboxylic acids. This was once a valuable method for structural determination.

R—≡—R →(O_3/O_2) ozonide intermediate →(H_2O) RCOOH + HOOCR

R—≡—R →($KMnO_4$) RCOOH + HOOCR

4.11 Metal Alkynides (Acetylides)

The terminal alkynes have a proton which is acidic and hence can be removed by a strong base. This forms an alkynide, which can react with a wide range of electrophiles.

R—≡—H ← acidic hydrogen relative to an alkane or vinyl hydrogen

$$\text{R—≡—H} \overset{\text{NaOH}}{\rightleftharpoons} \text{R—≡}^{-}\ \text{Na}^{+} + \text{H}_2\text{O}$$

$$\text{R—≡—H} \overset{\text{NaNH}_2}{\rightleftharpoons} \text{R—≡}^{-}\ \text{Na}^{+} + \text{NH}_3$$

To form salts a strong base is needed: NaH, $NaNH_2$, Na, RLi.

4.12 Acidity of Alkynes

$$\text{RH} + \bar{\text{B}} \longrightarrow \bar{\text{R}} + \text{BH}$$

$CH_3\bar{C}H_2$ $\quad$ $CH_2{=}\bar{C}H$ $\quad$ H—≡$^{-}$

stability of carbanion increases →

The acidity of RH is directly proportional to the stability of R^-. Why does the stability of R^- increase along the series below?

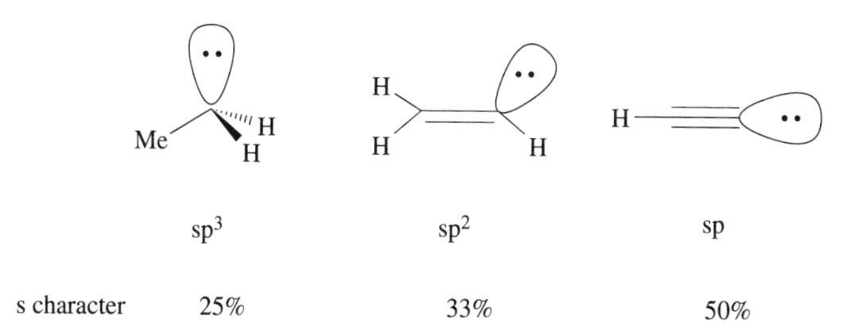

The more s character there is, the closer the electrons are to the nucleus and are therefore more stable.

4.13 Alkyne Dianions

If the base used is strong and is in excess, a second proton may be abstracted. The second anion formed is the more reactive one.

$$CH_3{-}C{\equiv}C{-}H \xrightarrow{\text{BuLi}} CH_3{-}C{\equiv}C^{-}\ Li^{+} \xrightarrow{\text{BuLi}}$$

$$Li^{+}\ \overset{-}{C}H_2{-}C{\equiv}C^{-}\ Li^{+} \xrightarrow[\text{1 eq.}]{\text{RI}} R{-}CH_2{-}C{\equiv}C^{-}\ Li^{+}$$

$$\xrightarrow{H_2O} R{-}CH_2{-}C{\equiv}C{-}H$$

$R{-}CH_2{-}C{\equiv}C^{-}\ Li^{+}$ is more stable than $Li^{+}\ \overset{-}{C}H_2{-}C{\equiv}C{-}R$

4.14 Relevant Questions

12(a), 38, 42(c), 47(c), 53(a and d), 75(b), 128(a), 135(f), plus parts of functional group interconversion questions.

5 ALCOHOLS AND THIOLS

The alcohols are central to organic chemistry; thus much of the material covered in this chapter is also covered elsewhere in the book. Reactions covered in other chapters are cross-referenced and not repeated here.

5.1 Structure of Alcohols and Thiols

Alcohols have a hydroxy group, —OH, attached to an alkyl group. The alkyl group may contain other functionality (aryl groups, double bond(s), halogens, other hydroxy groups and so on). Alcohols are classified as methyl, primary, secondary, or tertiary.

CH_3OH	RCH_2OH	R_2CHOH	R_3COH
methyl	primary	secondary	tertiary

R = alkyl, aryl group

The geometry of methanol is given below, along with a diagram showing how the two lone pairs are staggered between two adjacent carbon–hydrogen bonds.

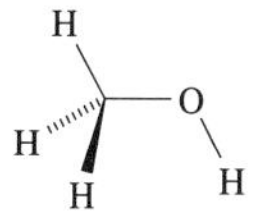

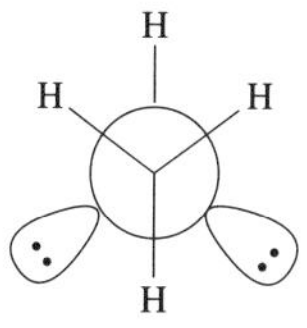

H H H O H

H H H H

Thiols (or mercaptans) are sulfur analogues of alcohols. In a thiol, a sulfhydryl (or mercapto) group, —SH, is bonded to an alkyl group with the same possibilities of substitution as in alcohols.

CH_3SH	RCH_2SH	R_2CHSH	R_3CSH
methyl	primary	secondary	tertiary

R = alkyl, aryl group

5.2 Physical Properties of Alcohols and Thiols

Alcohols, like water, form hydrogen bonds and for the simpler ones the boiling points are generally higher than that of water (see Streitweiser *et al.* 1992 for further details).

The infrared spectroscopy of alcohols is dominated by the stretching and vibrations of the —OH group. In non-polar solvents such as carbon tetrachloride, where there is little association between molecules, alcohols show an O–H stretching band at about 3640 cm^{-1}. As the concentration of alcohol is incresed this band is replaced by a broader band at about 3350 cm^{-1}. The hydrogen bonding weakens the O-H bond, and lowers the energy and hence the frequency of vibration.

5.3 Industrial Preparation

There three main industrial sources of alcohols; these are hydration of alkenes, fermentation and the oxo process.

5.4 Hydration of Alkenes

The alkenes obtained from the cracking of petroleum can be hydrated to form alcohols. Only those alcohols consistent with the application of Markovnikov's rule (see 3.17 to 3.22) can be obtained. The only primary alcohol accessible by this method is ethanol.

$$R_2C{=}CR_2 \xrightarrow{H_2SO_4} R_2CH{-}CR_2(OSO_3H) \xrightarrow{H_2O} R_2CH{-}CR_2(OH)$$

5.5 Fermentation

The fermentation of sugars by yeast is of enormous economical importance for the preparation of ethanol and certain other alcohols. The sugars come from a variety of sources, mostly molasses from sugar cane and starch obtained from various grains.

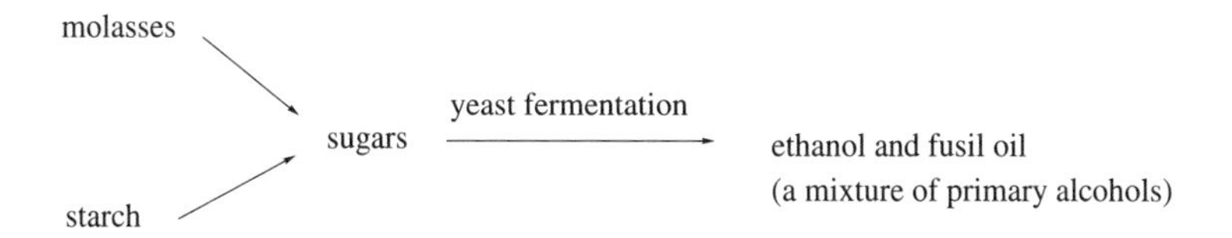

5.6 The Oxo Process

The oxo process allows the formation of primary alcohols. In the presence of octacarbonyldicobalt as a catalyst, alkenes react with carbon monoxide and hydrogen to yield aldehydes. These are then reduced by hydrogen and the catalyst to alcohols. The oxo process is essentially the addition of hydrogen and the formyl group across the double bond and is therefore known as hydroformylation.

$$R_2C{=}CR_2 \xrightarrow[Co_2(CO)_8]{CO,\ H_2,} H{-}CR_2{-}C(R)(R)CHO \xrightarrow[Co_2(CO)_8]{H_2} H{-}CR_2{-}C(R)(R)CH_2OH$$

5.7 Laboratory Preparation

There are many methods for the laboratory preparation of alcohols; of these only a few common or important methods are described here.

5.8 Hydrolysis of Alkyl Halides

See 8.12 to 8.19.

5.9 Oxymercuration–Demercuration

See 3.30.

5.10 Hydroboration–Oxidation

See 3.31.

5.11 Grignards with Carbonyl Compounds

See 8.10 and 10.37.

5.12 Organolithiums with Carbonyl Compounds

Organolithiums can react with carbonyl compounds in much the same way as Grignard reagents.

$$R_2C{=}O \xrightarrow{R^1Li} R_2C(OLi)R^1 \xrightarrow{H^+} R_2C(OH)R^1$$

5.13 Grignards with Epoxides

See 7.10.

5.14 Lithium aluminium Hydride with Epoxides

See 7.11.

5.15 Reduction of Carbonyls

See 10.10 to 10.12 and 10.75.

5.16 Conversion of Alcohols to Alkyl Halides Using Hydrogen Chloride and Zinc Chloride

See 8.3.

5.17 Conversion of Alcohols to Alkyl Halides Using Hydrochloric Acid

See 8.3.

5.18 Conversion of Alcohols to Alkyl Halides Using Thionyl Halides

See 8.4.

5.19 Conversion of Alcohols to Alkyl Halides Using Phosphorous Halides

See 8.4.

5.20 Alcohols and the Formation of Esters

See 10.59 to 10.61.

5.21 Alcohols and the Formation of Acetals

See 10.21 to 10.23.

5.22 Alcohols and the Formation of Alkyl Sulfonates

Alcohols will react with alkyl or aryl sulfonyl halides to produce alkyl or aryl sulfonates. The alkyl or aryl sulfonates are potent alkylating agents because the sulfonate anion is a reactive leaving group. The four common alkyl or aryl sulfonates are given in 3.3.

OH — $\xrightarrow{MeSO_2Cl}$ (pyridine, N) — OSO_2Me

5.23 Displacement Reactions of Alkyl Sulfonates

Alkyl or aryl sulfonates undergo displacement reactions to form alkenes, as in 3.9, or undergo nucleophilic substitution in a similar manner to alkyl halides as suggested in 8.11.

OSO_2Me base

OSO_2Me $CH_3CO_2^- K^+$ acetone O O Me

5.24 Formation of Alkenes from Alcohols

See 3.5, 3.6 and 3.9 to 3.11. Also see 3.7 for carbocation rearrangements.

5.25 Formation of Ethers from Alcohols, Reaction of Alkoxides with Alkyl Halides

See 6.1.

5.26 Oxidation of Alcohols

See 10.9 and 10.53.

5.27 Diols

Diols are usually prepared from the corresponding alkene by the net addition of two hydroxy groups to the double bond, see 7.6 and 3.25 to 3.27.

One of their main uses is as protecting groups for ketones and aldehydes, see 10.22. Certain diols undergo a rearrangement known as the pincol–pinecolone rearrangement, see 10.41 and 10.42. Dehydration of 1,4- and 1,5-diols often leads to the formation of cyclic ethers, particularly if one of the hydroxy groups is tertiary.

HO~~~~OH $\xrightarrow{H^+}$ (tetrahydrofuran, O)

5.28 Preparation of Thiols Using Sodium Hydrosulfide and Alkyl Halides

Thiols can be prepared from alkyl halides by displacement of the halide with a hydrogensulfide ion, HS^-, in ethanol.

$$\text{CH}_3(\text{CH}_2)_5\text{I} \xrightarrow{\text{NaSH, EtOH}} \text{CH}_3(\text{CH}_2)_6\text{SH} + \text{NaI}$$

A large excess of hydrosulfide is required because of the equilibrium below. Thiol anion produced is also a good nucleophile and can react with the alkyl halide to produce the corresponding sulfide.

$$HS^- + RSH \rightleftharpoons H_2S + RS^-$$

5.29 Preparation of Thiols Using Thiourea and Alkyl Halides

Thiourea has almost exclusively replaced hydrosulfide because of the problem above. Thiourea is soluble in water and alcohols and the sulfur is nucleophilic and readily takes part in S_N2 displacement reactions of alkyl halides. The product salt is readily hydrolised to the alkanethiol.

$$\text{~~~Br} \xrightarrow{H_2N-C(=S)-NH_2} \text{~~~S-C}(=\overset{+}{N}H_2)NH_2\ Br^-$$

$$\xrightarrow{NaOH/H_2O} \text{~~~SH} + H_2N-C(=O)-NH_2$$

5.30 Preparation of Thiols Using Grignards with Sulfur

Thiols can also be prepared by the reaction of Grignard reagents with sulfur.

MgBr $\xrightarrow{S_8}$ SMgBr

$\xrightarrow{HCl}$ SH + MgBrCl

5.31 Reactions of Thiols

The formation of sulfides, cleavage of sulfides, oxidation of sulfides to sulfoxides and sulfones, and the formation of sulfonium salts are dealt with in 6.1, 6.4, 6.6 and 6.7 respectively.

5.32 Formation of Disulfides

Thiols are readily oxidised to disulfides. The disulfide bond is weak and easily reduced to give the thiol. Mild oxidising agents such as iodine suffice for the oxidation. A common reducing agent for regeneration of the thiol is lithium in liquid ammonia.

2 SH $\xrightarrow{I_2}$ S–S

S–S $\xrightarrow[\text{NH}_3\text{, then H}^+]{\text{Li in liquid}}$ 2 SH

5.33 Dithiols

Dithiols are often seen as protecting groups for aldehydes and ketones, see 10.23. The resulting thioacetals (1,3-dithianes) are resistant to acid hydrolysis, unlike their acetal analogues.

1,3-Dithianes are also used to form nucleophiles from aldehyde, resulting in another method for introducing the carbonyl group into a molecule.

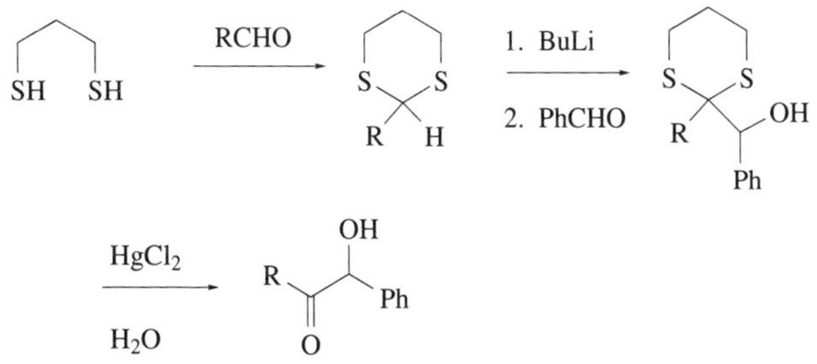

5.34 Relevant Questions

34(e), 38, 40, 81(a), 137(d), 149, plus parts of functional group interconversion questions.

6 ETHERS AND SULFIDES

6.1 The Williamson Synthesis of Ethers and Sulfides

Ethers can be prepared by the action of an alkoxide on an alkyl halide. The same reaction can be used to form a sulfide.

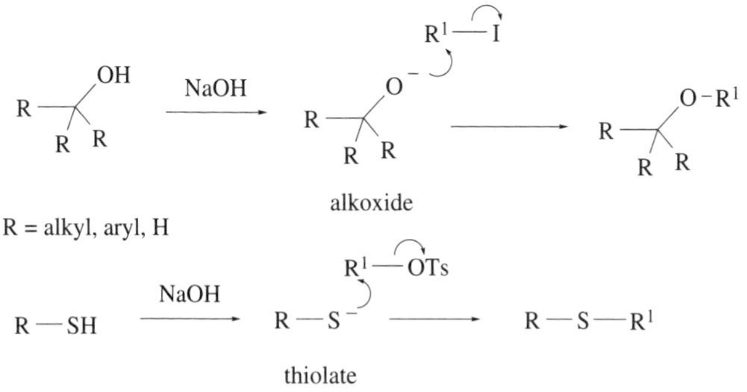

R = alkyl, aryl, H

alkoxide

thiolate

6.2 Synthesis of Ethers by Alkoxymercuration of Alkenes

Ethers can also be formed by alkoxymercuration of alkenes.

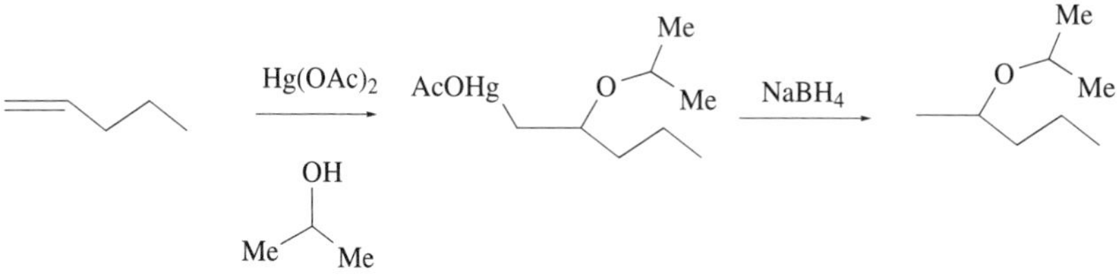

6.3 Acid Cleavage of Ethers

Ethers can be cleaved by strong acids such as hydrogen bromide or iodide. The alcohol formed initially is converted to the corresponding alkyl halide.

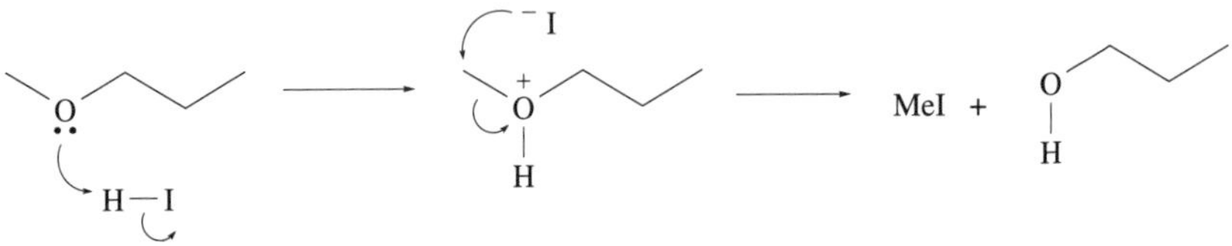

With tertiary ethers, carbocations are involved and the reactions tend to be much more complex. Heating such ethers with strong acid often leads to elimination.

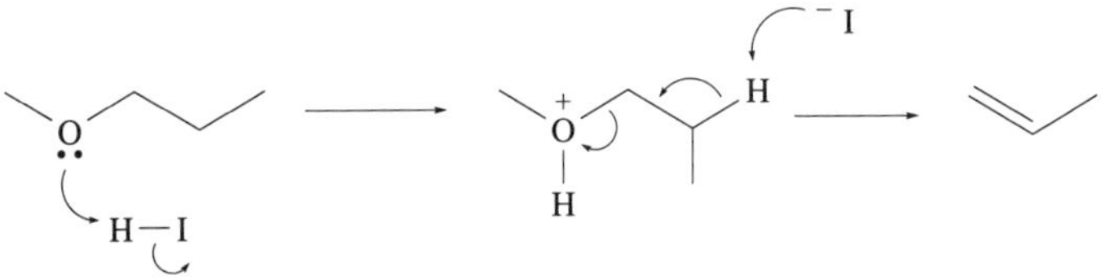

6.4 Cleavage of Sulfides Using Raney Nickel

See 10.24.

6.5 Lability of Benzyl Ethers and their Use as Protecting Groups

Benzyl ethers can be removed by hydrogenation. They can be used as a protecting group for alcohols and be converted back to alcohols selectively when required.

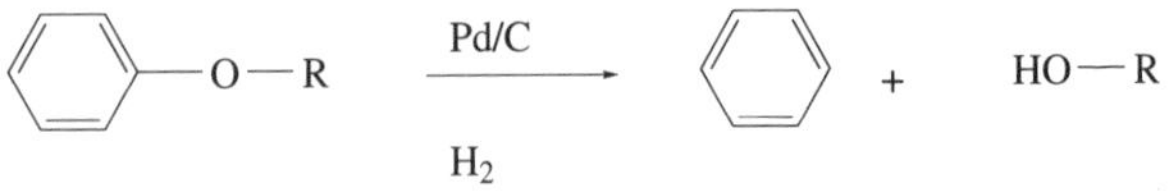

6.6 Oxidation of Sulfides to Sulfoxides and Sulfones

Sulfides are easily oxidised, initially to sulfoxides and then by further oxidation to sulfones. A convenient oxidant, which converts sulfides to sulfoxides without the production of the sulfone is sodium periodate.

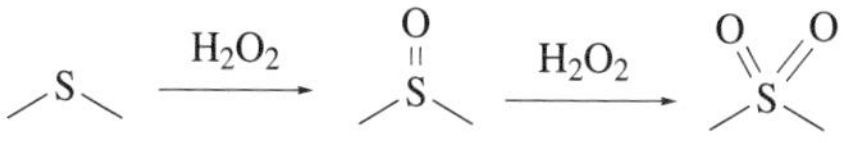

H_2O_2 H_2O_2

dimethyl sulfoxide dimethyl sulfone

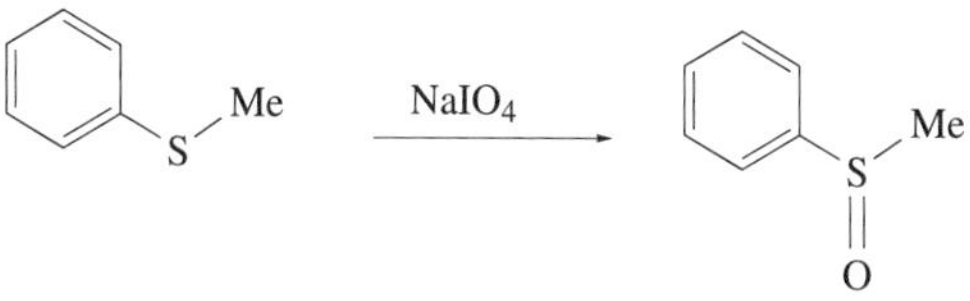

Me $NaIO_4$ Me

Dimethyl sulfoxide is an important solvent in the class of aprotic solvents. It is cheap, colourless and odourless. It is prepared industrially by the NO_2-catalysed air oxidation of dimethyl sulfoxide.

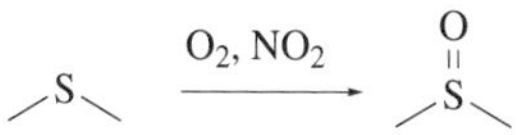

O_2, NO_2

dimethyl sulfoxide

6.7 Basicity of Ethers and Sulfides: Formation of Oxonium and Sulfonium Salts

Ethers and sulfides can form oxonium and sulfonium salts respectively. The oxonium ion is thought to be instrumental in the hydrolysis of acetals, see 10.22 to 10.23.

oxonium ion 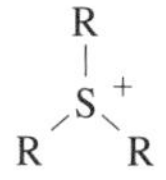 sulfonium ion

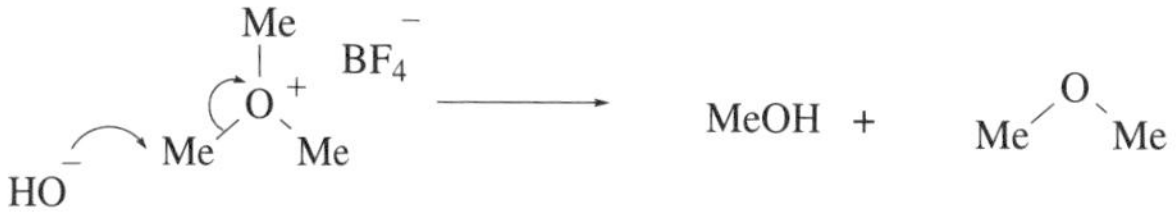

$HO^- + Me_3O^+ \, BF_4^- \longrightarrow$ MeOH + $MeOMe$

6.8 Relevant Questions

106(b), plus parts of functional group interconversion questions.

7 EPOXIDES

7.1 General Properties of Epoxides

The epoxides or oxiranes are a group of three membered cyclic ethers that behave differently from all other cyclic and acyclic ethers. Their high reactivity is due to the strain in the ring, where there is a large departure from the tetrahedral angle.

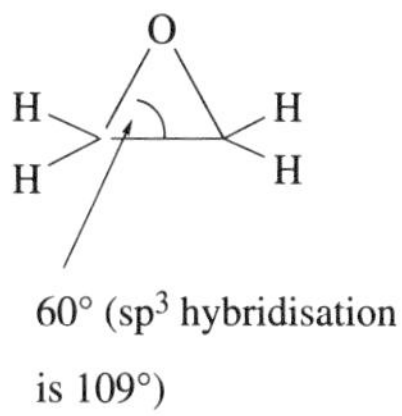

60° (sp^3 hybridisation is 109°)

7.2 Industrial Preparation of Epoxides

Ethylene oxide is prepared by passing ethene and air over a silver catalyst at 300 °C. This is the simplest epoxide. The procedure is not so useful for forming other epoxides, and is practically useless in the laboratory.

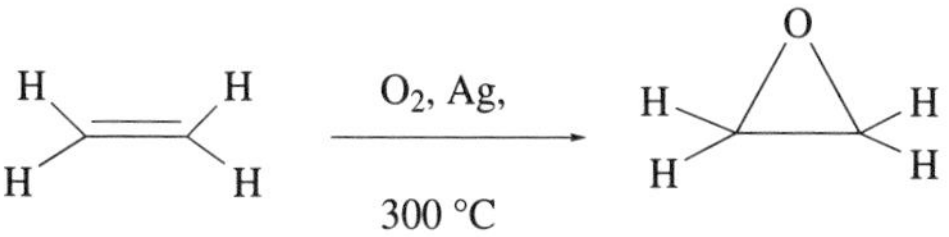

O_2, Ag, 300 °C

7.3 Laboratory Preparation from Halohydrins

If halohydrins are treated with a base an epoxide is formed. This is an **intramolecular** ("within the same molecule", as opposed to **intermolecu-**

lar, "between different molecules") Williamson synthesis (see 6.1). Note the *anti*periplanar stereoelectronic requirement.

NaOH

H_2O

intramolecular substitution

+ Br^-

7.4 Laboratory Preparation by Oxidation of Double Bonds with Peroxy Acids

An alkene can be oxidised using a peroxy acid (*e.g. meta*-chloroperbenzoic acid (MCPBA)). The addition of oxygen is always *syn*.

peroxy acid

RCO_2OH

an epoxide

7.5 Reactions of Epoxides

Reactions of epoxides are dominated by ring opening because of the release of ring strain on opening to an alkyl chain.

7.6 Ring Opening of Epoxides by Acid

Epoxides can be hydrolysed to the diol in an *anti* dihydroxylation fashion. Note that if a cyclic alkene is dihydroxylated in this maner, a *trans* diol is produced.

trans diol

7.7 Regiochemistry of Epoxide Ring Opening

If HCl in dry ether is used to ring-open an epoxide then the nucleophile attacks at the more substituted carbon (carbocation stability).

CH_3 O: H H^+ HCl ether Cl⁻ CH_3 δ+ OH δ+ H Cl CH_3 OH H

tertiary carbocation being formed, as opposed to the secondary carbocation that would be formed at the other carbon

7.8 Ring Opening of Epoxides

Epoxides can be attacked by nucleophiles. Some of the types of nucleophiles which will attack an epoxide follow. Notice that in all cases an alcohol is formed and that the attack is in an *anti*periplanar fashion.

7.9 Ring Opening of Epoxides by Hydroxides and Alkoxides

O RO^- O^- RO H^+ OH RO

R = alkyl, aryl or H

7.10 Ring Opening of Epoxides by Grignard Reagents

O R—MgX OMgX R H^+ OH R

R = alkyl or aryl, X = halogen

7.11 Ring Opening of Epoxides by Lithium Aluminium Hydride

"H^-" O $\xrightarrow{LiAlH_4}$ H O$^-$ $\xrightarrow{H^+}$ H OH

7.12 Ring Opening of Epoxides by Ammonia and Amines

This gives a route to amino alcohols.

$R_2\ddot{N}H$ O $\longrightarrow$ R_2N OH

7.13 Relevant Questions

10(b), 14(c), 22(b), 56(b), 71(d), 137(b)(iii), plus parts of functional group interconversion questions.

8 ALKYL HALIDES

The alkyl halides are a class of compounds where a hydrogen atom in an alkane has been replaced by a halogen atom. It is possible to replace more than one hydrogen by halogens to form di, tri and other multi-halogenated compounds. This chapter deals only with the mono halogenated alkanes.

Like the alkanes, structural isomers exist once a certain chain length is reached; for alkyl halides this length is three carbons.

CH_3X X = Cl, Br, I
C_2H_5X

The alkyl halide C_3H_7X has two structural isomers:

$CH_3CH_2CH_2X$	CH_3CHXCH_3
primary	secondary

The alkyl halide C_4H_7X has four structural isomers:

X	X	X	X
n-butylhalide	*s*-butylhalide	*t*-butylhalide	*i*-butylhalide

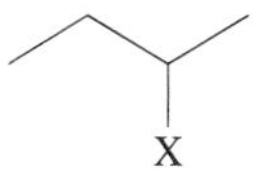

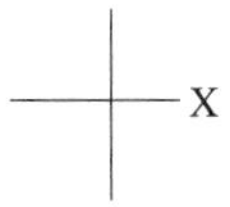

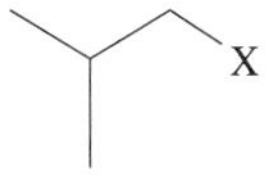

8.1 Industrial Preparation of Alkyl Halides

The main industrial preparation of alkyl halides is by free-radical halogenation. When an alkane is treated with Cl_2 or Br_2 in the presence of heat or light, a mixture of alkyl halides is formed. These can be separated on an industrial scale. Methane, for example, produces four alkyl halides.

$$CH_4 \xrightarrow[\text{or light}]{Cl_2,\ \text{heat}} CH_3Cl + CH_2Cl_2 + CHCl_3 + CCl_4$$

8.2 Laboratory Preparation of Alkyl Halides

Laboratory preparation of alkyl halides mainly relies on the conversion of a functional group, such as an alcohol or double bond, to a halide.

8.3 Alkyl Halides from Alcohols by the Use of Hydrogen Halides

The concentrated aqueous hydrogen halides convert the hydroxyl into a good leaving group by protonation. The protonated hydroxyl group then either leaves as water to give a carbocation (S_N1 mechanism), or is easily displaced by the halide anion (S_N2 mechanism). The S_N1 and S_N2 mechanisms are discussed in Sections 8.12 to 8.19 of this chapter.

$$ROH \overset{HX}{\rightleftharpoons} RX + H_2O$$

HX is concentrated or gas

Zinc chloride can also be used in conjunction with concentrated hydrochloric acid; the zinc chloride facilitates the reaction as it is a powerful Lewis acid that serves the same purpose as does the proton in coordinating with the hydroxy oxygen.

Tertiary is easier than secondary which is easier than primary (need forcing conditions, i.e. high temperature, very concentrated acid) because the positive charge built up at the reacting centre is stabilised by the tertiary group better than by either a secondary or a primary group.

8.4 Alkyl Halides from Alcohols by the Use of Thionyl and Phosphorus Halides

Thionyl halides, especially thionyl chloride, are often used to convert hydroxyl groups to halides. They have the advantage of being relatively mild and specific. The side products are gases; this has an added advantage in terms of purification of the products.

$$ROH \underset{}{\overset{SOX_2}{\rightleftharpoons}} RX + SO_2 + HX$$

X = Cl or Br

There are other methods, which include the use of PCl_3, PI_3 and PBr_3. Reaction of these with an alcohol produces a phosphite ester and hydrogen halide. The phosphite ester reacts with the hydrogen halide to give the alkyl halide.

$$(CH_3)_2CHCH_2CH_2OH \xrightarrow{PBr_3} [(CH_3)_2CHCH_2CH_2O]_3P + HBr$$

$$\longrightarrow (CH_3)_2CHCH_2CH_2Br$$

The advantage of these reagents over the hydrogen halides is that milder reaction conditions are required and carbocation rearrangements are less important.

A point to note is that the chemistry of alkyl halides is sometimes very similar to that of toluene sulfonyl derivatives of alcohols (tosylates), see 5.22 and 5.23.

$$ROH \xrightarrow{TsCl} ROTs \qquad Ts = -SO_2-C_6H_4-Me$$

8.5 Alkyl Halides by Allylic or Benzylic Halogenation

Allylic or benzylic hydrogens can be replaced by bromine using *N*-bromosuccinimide (NBS). See also 3.32 and March (1992) for mechanism.

R R NBS R R
R R R R R
R R
H Br

NBS

R R
R R
H Br

NBS = N—Br (with two C=O groups, O and O)

8.6 Alkyl Halides by Halogenation of Alkenes

This is covered in 3.23 and 3.24.

8.7 Alkyl Iodides from Alkyl Bromides or Chlorides

The interconversion of one alkyl halide to another, using a salt source of the new halide, is quite common, especially in the formation of alkyl iodides. The reaction is heavily dependent on the solvent used in the reaction.

$$\text{RBr or RCl} \xrightarrow{\text{NaI}} \text{RI} + \text{NaCl or NaBr}$$

NaI is highly soluble in acetone. NaCl and NaBr are insoluble. As NaCl and NaBr are formed they drop out of solution (are precipitated) pushing the equilibrium to the right-hand side towards RI. This also works for ROTs, see 5.22 and 3.3.

This route allows the formation of "anti-Markovnikov" iodoalkanes.

$$RCH{=}CH_2 \longrightarrow RCH_2CH_2Br \longrightarrow RCH_2CH_2I$$

8.8 Bromo- or Iodoalkanes from Silver Salts of Carboxylic Acids

$$RCO_2H \longrightarrow RCO_2Ag \xrightarrow[\text{in } CCl_4]{Br_2 \text{ or } I_2} RCO_2Br + AgBr$$

$$RCO_2Br \longrightarrow RBr + CO_2$$

Mercury or lead salts can also be used in this reaction.

8.9 Reactions of Alkyl Halides: Elimination to Give Alkenes

Alkyl halides can be converted to alkenes. This is covered in 3.3 and 3.4.

$$RX \xrightarrow[E2]{\text{base}} \text{alkene}$$

8.10 Reactions of Alkyl Halides: Formation of Grignards and Lithium Species

An alkyl halide will in general react in a dry atmosphere with magnesium to form what is effectively an alkyl anion. These compounds are known as Grignard reagents. Reaction of lithium with an alkyl halide, also in a dry and inert atmosphere, produces a species known as an alkyl lithium. Both Grignards and alkyl lithiums will react with electrophiles.

$$RX \xrightarrow[Li]{Mg \text{ or}} RMgX \text{ or } RLi \xrightarrow[\text{"E"}]{\text{electrophiles}} RE$$

This does not occur for toluene sulfonyl derivatives (tosylates).

8.11 Reactions of Alkyl Halides: Nucleophilic Substitution Reactions

$$RX \xrightarrow{\text{nucleophile Y}} RY$$

The carbon–halogen bond in an alkyl halide is polarised, such that there is more electron density on the halogen than on the carbon, as shown below when X is a halogen. If X, known as the **leaving group**, is stable as X^-, as it is in the case of the halogens, it can be displaced by another group, Y. The group Y is known as a nucleophile and carries either a lone pair of electrons or a negative charge (in other words, excess electron density). The carbon attached to X is said to be electrophilic.

δ^- X leaving group, X^- is stable

δ^+

the central carbon is electrophilic (δ+) due to -I effect of X

$$Y^- + R\text{-}X \longrightarrow Y\text{-}R + X^-$$

8.12 The S_N1 (Substitution Nucleophilic Unimolecular) Mechanism

For the reaction

$$HO^- + Me_3CBr \xrightarrow[\text{aqueous}]{\text{dil. NaOH}} Me_3COH + Br^-$$

the fact is that the kinetics are independent of $[HO^-]$, where [] means "concentration of". It is a first-order reaction where the rate $= K[Me_3CBr]$, where K is the rate constant. The theory is that it is a unimolecular mechanism (S_N1) where only the Me_3CBr is involved in the rate-limiting step (RLS).

$$Me_3C\text{—}Br \underset{\text{RLS}}{\overset{\text{slow}}{\rightleftharpoons}} Me_3C^+ + Br^-$$

$$Me_3C^+ + HO^- \xrightarrow{\text{fast}} Me_3C\text{—}OH$$

two steps

It is found that the alkyl halides show a decrease in their ability to react through the S_N1 mechanism according to the scheme below.

$Me_3X > Me_2CHX > MeCH_2X > MeX$

decreasing reactivity by S_N1 mechanism because there is a decrease in carbocation-like transition state as we go along the series tertiary to primary carbocation (there is an increase in $\Delta G^{\#}$)

Looking a little closer at the RLS

planar, trigonal sp^2 hybridised carbocation

we find that the transition state has some carbocation character. Whatever makes the carbocation stable will therefore make the transition state more stable.

There is some evidence that S_N1 reactions are accelerated if the molecule is under strain to start with, due to a, b and c being bulky. The explanation is that in the transition state there is some lengthening of bonds and the groups start to pull away from each other slightly. Some of the original strain is relieved and the reaction is accelerated.

8.13 The Stereochemistry of the S_N1 Mechanism

Reactions that go *via* the S_N1 mechanism are non-stereospecific (racemisation occurs).

$$\underset{S\ \text{enantiomer}}{\overset{*}{\text{PhCHClMe}}} \xrightarrow{^{-}\text{OH}} \underset{56\%\ R\ \text{enantiomer},\ 44\%\ S\ \text{enantiomer}}{\text{PhCH(OH)Me}}$$

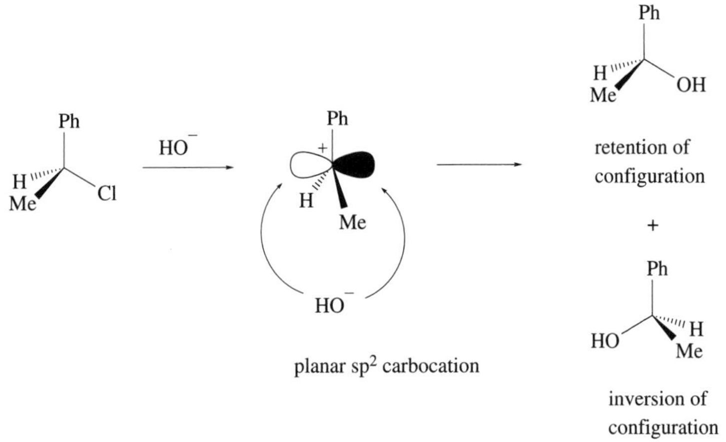
Ph
H
Me
Cl
HO⁻
Ph
+
H
Me
HO⁻
planar sp² carbocation
Ph
H
Me
OH
retention of configuration
+
Ph
HO
H
Me
inversion of configuration

The HO⁻ can attack either lobe of the 2p orbital. This would be completely correct if we obtained a 1:1 mixture of the products. To complete the picture we must consider that the leaving group Cl⁻ hinders the attack of the HO⁻ from the front and we see slightly more attack from behind. Thus we get slightly more inversion than retention.

8.14 The Competing E1 Elimination

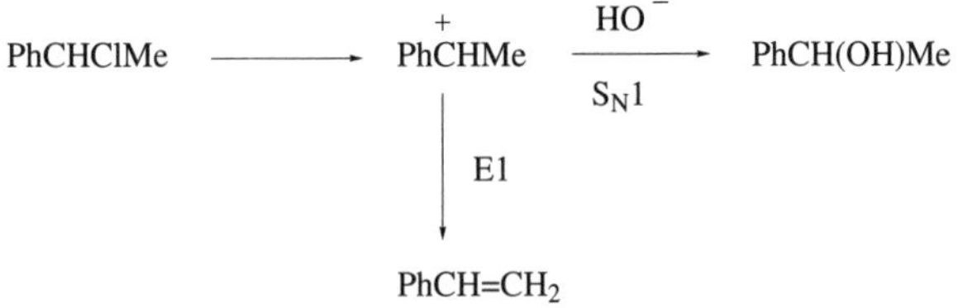
PhCHClMe
+
PhCHMe
HO⁻
S_N1
PhCH(OH)Me
E1
PhCH=CH₂

There will always be competition from E1 if you are trying to carry out an S_N1 reaction. The more powerful the nucleophile the more likely it is to attack the carbocation before the carbocation loses a proton to form the alkene, see 3.5.

8.15 The S_N2 (Substitution Nucleophilic Bimolecular) Mechanism

For the reaction

$$HO^- + MeBr \xrightarrow[\text{aqueous}]{\text{dil. NaOH}} MeOH + Br^-$$

the fact is that the kinetics are dependent on both $[CH_3Br]$ and $[HO^-]$. It is a second-order reaction where the rate $= K[CH_3Br][HO^-]$. The theory is that it is a bimolecular reaction in which both CH_3Br and HO^- are involved in the rate-limiting step.

$$HO^- \; CH_3{-}Br \xrightarrow{\text{RLS}} MeOH + Br^- \qquad \text{single step}$$

8.16 Details of the S_N2 Mechanism

Stereochemistry: there is stereospecific inversion of configuration.

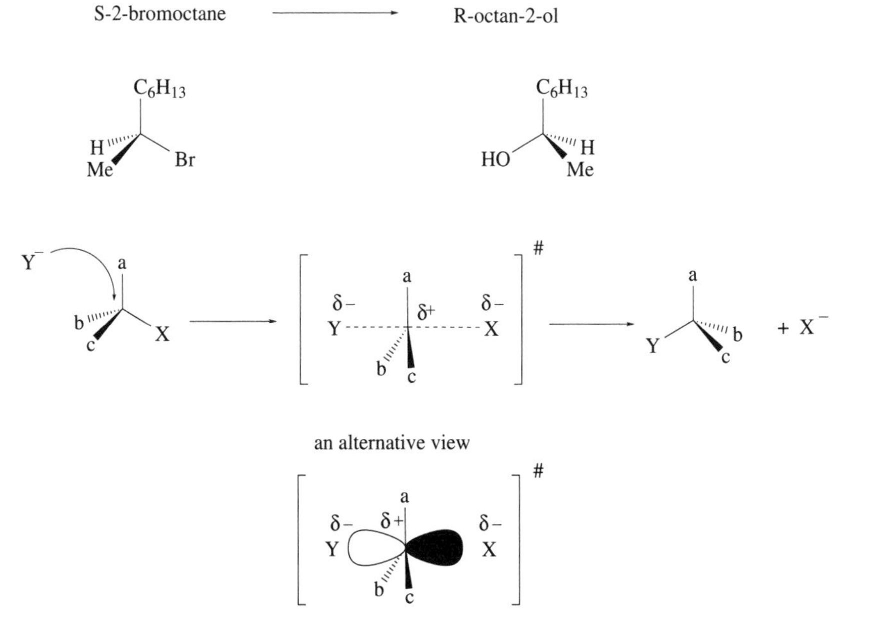

8.17 Influence of R on the Reactivity of RX in the S_N2 Mechanism

Fact:

MeX $MeCH_2X$ Me_2CHX Me_3CX

decreasing reactivity by S_N2 because there is increasing steric strain in S_N2 transition state, this causes an increase in $\Delta G^{\#}$

Therefore there is an increased crowding in the S_N2 transition state. If a, b and c are large, below, crowding will cause steric strain (overlap of van der Waals radii).

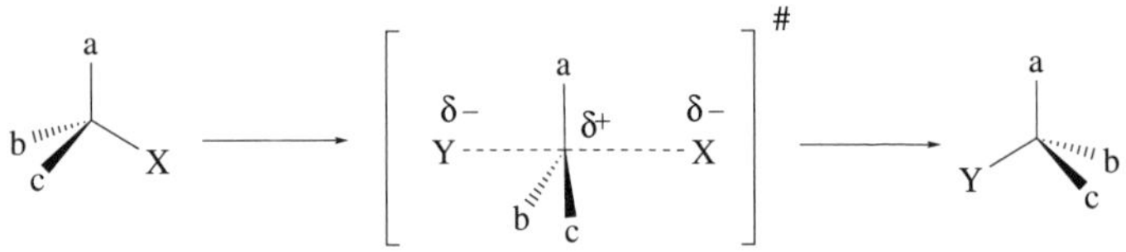

number of groups at central carbon: 4, 5, 4

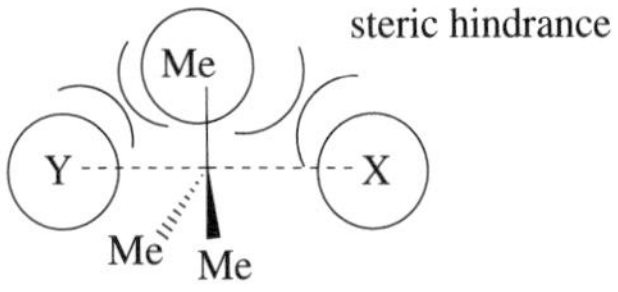

steric hindrance

8.18 Competition between S_N2 and E2 Mechanisms

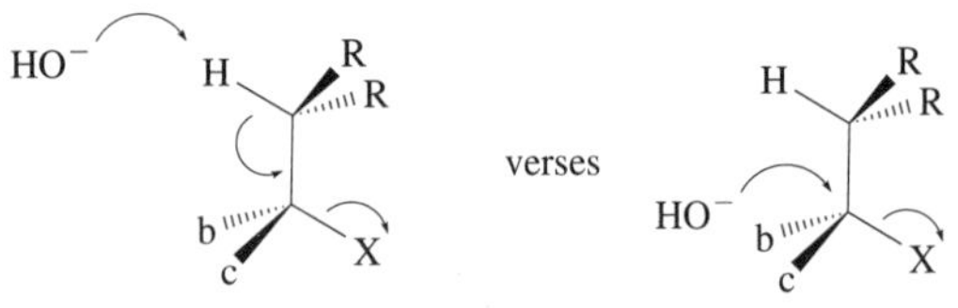

E2 verses S_N2

If b and c are methyl groups or something bigger, the S_N2 mechanism will be very slow and the E2 reaction mechanism will dominate.

8.19 Summary of S_N1, S_N2, E1 and E2 Mechanisms

S_N1 reactivity: MeX $<$ 1° RX $<$ 2° RX $<$ 3° RX.
S_N2 reactivity: MeX $>$ 1° RX $>$ 2° RX $>$ 3° RX.
MeX, 1° RX: as S_N2 is fast and S_N1 is slow these will choose to react by the S_N2 mechanism because this has a low $\Delta G\#$.
3° RX: as S_N1 is fast and S_N2 is slow these will choose to react by the S_N1 mechanism because here $\Delta G\#$ is low.
2° RX: both the S_N1 and S_N2 mechanisms are moderately easy or hard. They are borderline S_N1/S_N2.

In these cases the observed rate for nucleophilic substitution is (rate for S_N1) + (rate for S_N2).

8.20 Allyl Halides

allyl X, CH_2=$CHCH_2X$

benzyl X, C_6H_5—CH_2X

In these molecules both S_N1 and S_N2 mechanisms are fast. The observed rate is therefore high, *i.e.* there is high reactivity. With regard to the S_N1 reaction the explanation is simple as the cation produced is resonance-stabilised. With S_N2, it is the positive charge build-up in the t-s which is stabilised.

8.21 Vinyl and Phenyl Halides

vinyl X, CH_2=CHX

phenyl X C_6H_5—X

In this case the S_N1 and S_N2 pathways are very slow and unreactive. Again, with regard to the S_N1 reaction this is easy to understand since the cation formed is very unstable because the carbon atom has a lot of s character and so it is unwilling to give up electrons. For the S_N2 case a different mechanism operates, see 11.66.

8.22 Neopentyl Halides

Me_3CCH_2X: the S_N1 and S_N2 mechanisms are both slow and have low reactivity. The S_N1 is slow because a 1° carbocation must be formed. The S_N2 is slow because the nucleophile has to squeeze itself in to get at the carbon (C_α) and so it is difficult to get an S_N2 reaction to occur due to steric hindrance created by the three methyl groups attached to C_β.

$$Me_3CCH_2X \longrightarrow Me_3C\overset{+}{C}H_2 \quad \text{primary cation}$$

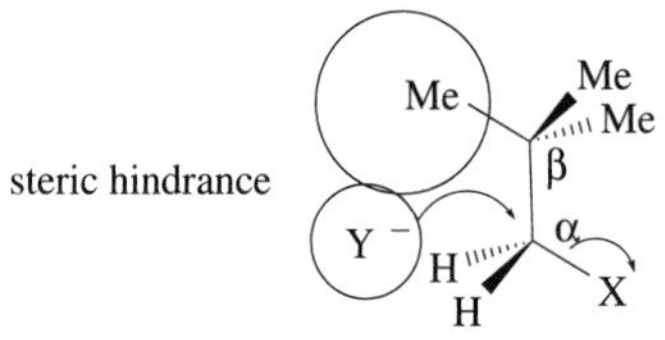

steric hindrance

Me Me Me β α Y⁻ H H X

8.23 Relevant Questions

50, 52, 89, 115(a), 117(a), 131(a and c), 137(c), 143(c), 147, 164, 167, plus parts of functional group interconversion questions.

9 AMINES

9.1 General Properties of Amines

There are primary, secondary and tertiary amines.

RNH_2	R_2NH	R_3N	
primary	secondary	tertiary	R = alkyl, aryl group

9.2 The Lone Pair in Amines

Nitrogen is less electronegative than oxygen and is therefore more willing to donate its pair of electrons.

The dominant reaction of amines is their ability to be protonated.

$$R_3N + H_2O \rightleftharpoons R_3\overset{+}{N}{-}H + HO^-$$

R = alkyl, aryl or hydrogen

$$K_b = \frac{[R_3\overset{+}{N}H][HO^-]}{[R_3N]}$$

The more alkyl groups attached to the nitrogen, the more stable the protonated species should be. However, the basicity does not only depend on the inductive (+I) effect of the alkyl groups. It also depends on the extent of solvation.

	K_b
CH_3NH_2	4.4×10^{-4}
$(CH_3)_2NH$	5.1×10^{-4}
$(CH_3)_3N$	5.3×10^{-5}

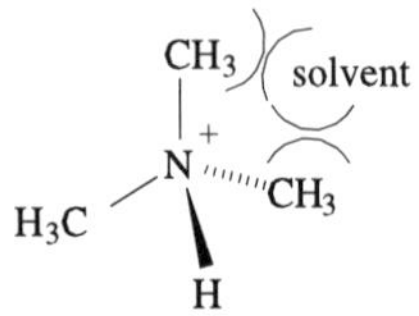

CH_3 solvent N^+ CH_3 H_3C H

Steric hindrance begins to prevent solvation. The inductive effect pushing the equilibrium to the right is being counterbalanced by this steric effect.

9.3 Hydrogen Bonding in Amines

Hydrogen bonding is the phenomenon whereby a weak electrostatic bond is formed, between hydrogen and other atoms (O, S, N, *etc.*) in a molecule.

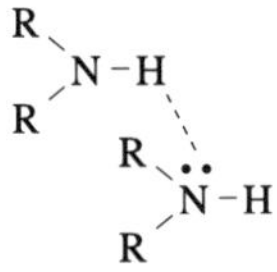

$R_2N{-}H \cdots :NR_2{-}H$

	n-C_5H_{12}	n-$C_5H_{11}OH$	n-$C_5H_{11}NH_2$
bp	36 °C	137 °C	78 °C

9.4 Stereochemistry of Amines

Amines have a distorted tetrahedral shape. Enantiomers interconvert rapidly *via* the sp^2 hybridised molecule.

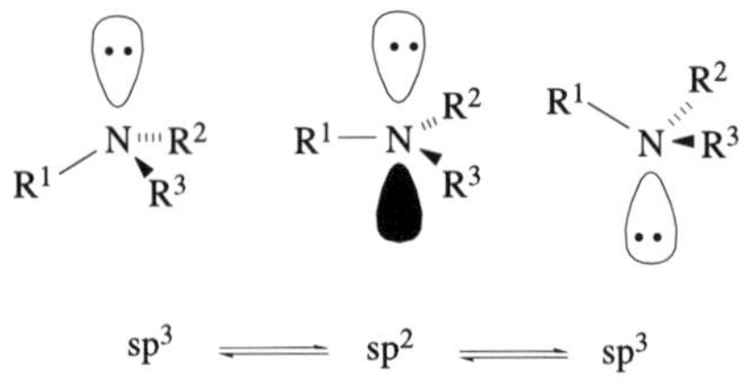

$sp^3 \rightleftharpoons sp^2 \rightleftharpoons sp^3$

If the groups R^1, R^2 and R^3 are connected then the enantiomers cannot interconvert, as in Troger's base below.

Me Me N N Troger's base

The diastereomers of Troger's base can be separated by chromatography. Some quaternery (4°) ammonium salts are also separable.

$$R^1R^2R^3N \xrightarrow[\text{quarternisation}]{R^4Br} [R^1R^2R^3R^4N^+]\ Br^-$$

a resolvable salt

9.5 Synthesis of Amines from an Alkyl Halide and Ammonia

Primary, secondary and tertiary amines can be made by the reaction of an alkyl halide with ammonia. This is in general only useful as an industrial synthetic method.

$$RBr \xrightarrow{NH_3} R\overset{+}{N}H_3\ Br^- \overset{NH_3}{\rightleftharpoons} RNH_2 + \overset{+}{N}H_4$$

$$RNH_2 \xrightarrow{RBr} R_2\overset{+}{N}H_2\ Br^-$$

$$R_2\overset{+}{N}H_2\ Br^- \xrightarrow[\text{2. NaOH}]{\text{1. repeat } NH_3,\ RBr} R_4NOH + RNH_2 + R_2NH + R_3N$$

$$R_4NOH + RNH_2 + R_2NH + R_3N \xrightarrow{\text{separate}} \text{various amines}$$

9.6 The Hinsberg Method for Separating Primary, Secondary and Tertiary Amines

The mixture of amines is reacted with *p*-toluene sulfonyl chloride and NaOH; the general scheme below is then followed. The amines are recovered on hydrolysis and basicification.

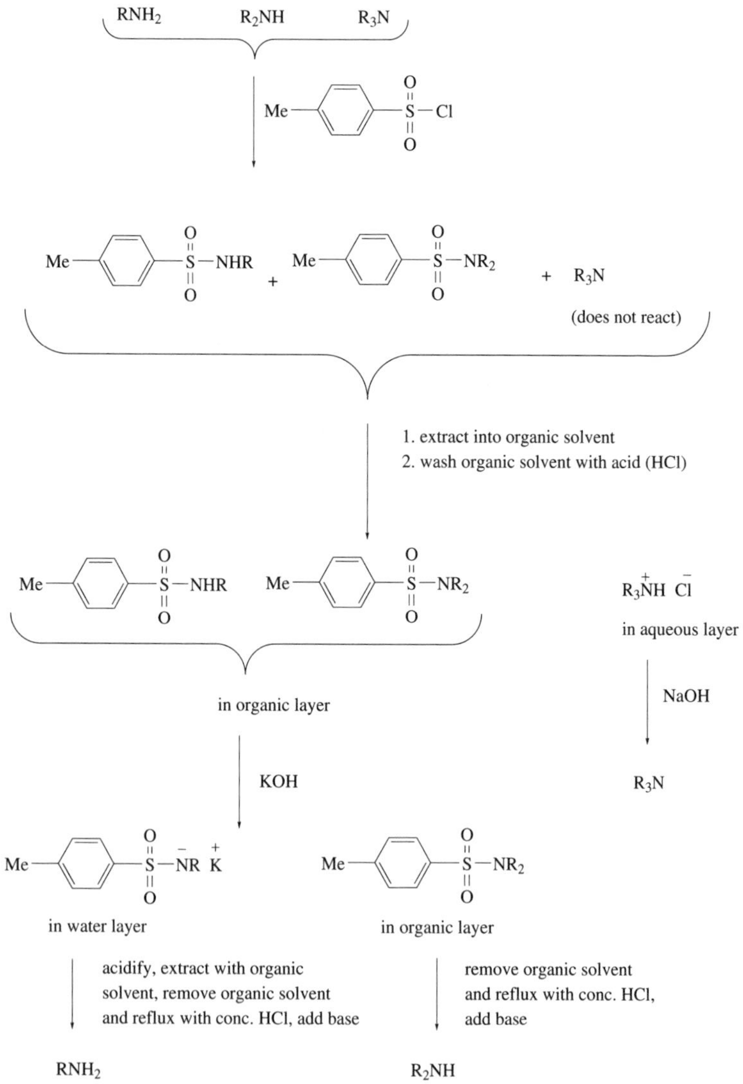

RNH_2 R_2NH R_3N

$Me-C_6H_4-SO_2-Cl$

$Me-C_6H_4-SO_2-NHR$ + $Me-C_6H_4-SO_2-NR_2$ + R_3N (does not react)

1. extract into organic solvent
2. wash organic solvent with acid (HCl)

$Me-C_6H_4-SO_2-NHR$ $Me-C_6H_4-SO_2-NR_2$

in organic layer

$R_3\overset{+}{N}H\ \overset{-}{Cl}$

in aqueous layer

NaOH

R_3N

KOH

$Me-C_6H_4-SO_2-\overset{-}{N}R\ \overset{+}{K}$

in water layer

acidify, extract with organic solvent, remove organic solvent and reflux with conc. HCl, add base

RNH_2

$Me-C_6H_4-SO_2-NR_2$

in organic layer

remove organic solvent and reflux with conc. HCl, add base

R_2NH

9.7 Gabriel Synthesis of Amines

This is a method whereby a source of nitrogen as a strong nucleophile is reacted with an alkyl halide. The original nitrogen is then released from the phthalimide by reaction with hydrazine.

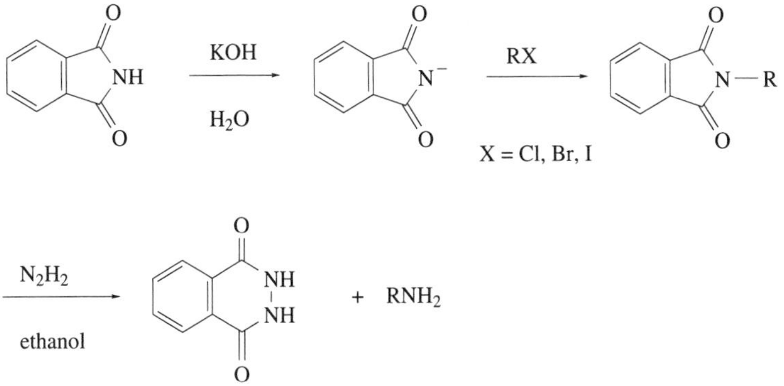

Phthalimide (NH) $\xrightarrow[H_2O]{KOH}$ phthalimide anion (N^-) $\xrightarrow{RX}$ N-alkylphthalimide (N—R)

X = Cl, Br, I

$\xrightarrow[\text{ethanol}]{N_2H_2}$ phthalhydrazide (NH—NH) + RNH_2

very insoluble in ethanol

9.8 The Curtius and Hofmann Reactions

See 10.81.

9.9 Amines by Reductive Methods

Some compounds containing nitrogen can be reduced using various reagents to amines.

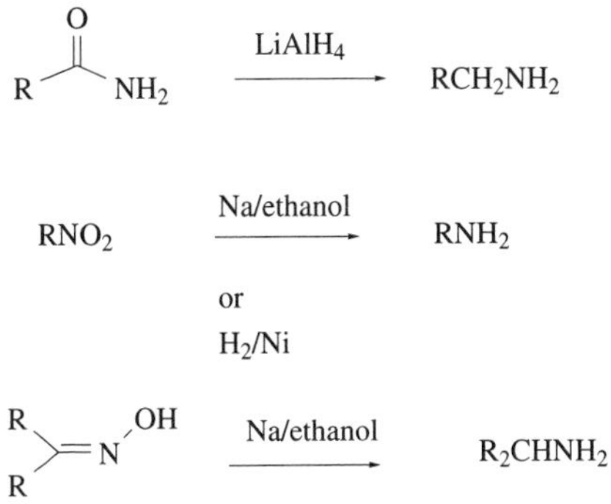

$$RC(=O)NH_2 \xrightarrow{LiAlH_4} RCH_2NH_2$$

$$RNO_2 \xrightarrow[\text{or } H_2/Ni]{\text{Na/ethanol}} RNH_2$$

$$R_2C{=}N{-}OH \xrightarrow{\text{Na/ethanol}} R_2CHNH_2$$

9.10 The Ritter Reaction

See 10.31.

9.11 Synthesis of Secondary Amines by Reduction of Imines

Imines can generally be formed from aldehydes and ketones with relative ease. The reduction of the carbon–nitrogen double bond results in a secondary amine.

$$R_2C{=}O \xrightarrow[\text{ethanol, } \textit{in situ} \text{ reduction } (H_2/Pd \text{ or } NaBH_3CN)]{RNH_2} R_2CHNHR$$

$$R_2C{=}O \rightleftharpoons R_2C{=}NR \rightarrow R_2CHNHR$$

9.12 Synthesis of Secondary Amines by Nitrosation of Dialkyl Aniline

This is a method for making secondary amines where one of the groups attached to the amine is a phenyl (or phenyl derivative). More importantly, further manipulation of the phenyl containing amine results in a secondary dialkyl amine being formed, as illustrated in the scheme below.

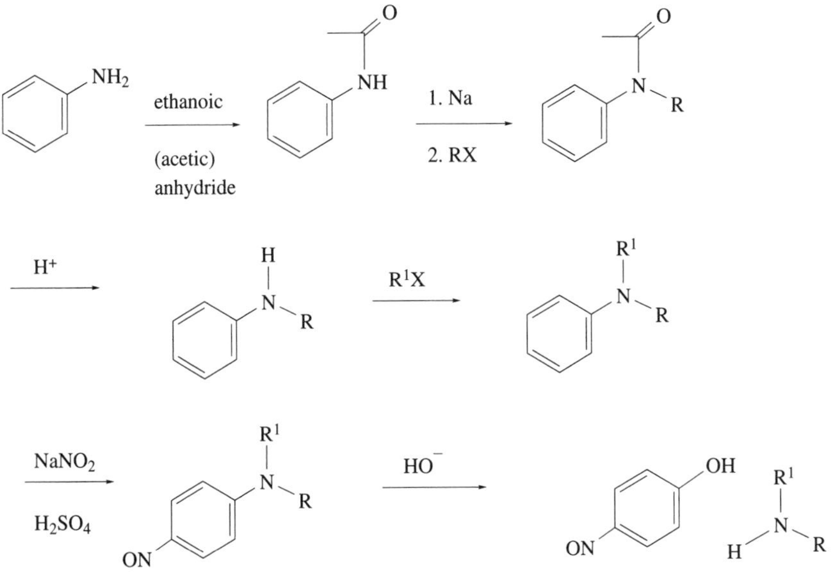

ethanoic (acetic) anhydride

1. Na 2. RX

H^+

R^1X

$NaNO_2$ H_2SO_4

HO^-

nitroso compound

9.13 Synthesis of Secondary Amines from Chloro Acetonitrile with Grignard Reagents

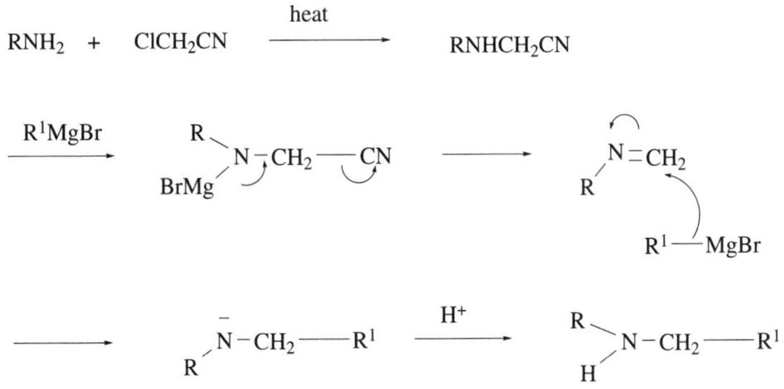

$$RNH_2 + ClCH_2CN \xrightarrow{\text{heat}} RNHCH_2CN$$

R^1MgBr

R^1—MgBr

H^+

9.14 Reactions of Amines: Formation of Salts

The formation of salts by amines (which are bases) is an obvious reaction and often occurs as a side reaction.

$$RNH_2 \xrightarrow{HX} R\overset{+}{N}H_3\ \overset{-}{X}$$

9.15 Reactions of Amines: Acylation

Acylation of an amine results in the formation of an amide. This is a useful reaction if it is necessary to protect the amine from, or prevent it participating in, other reactions during a synthetic sequence. The amide can often be converted back to the amine by hydrolysis.

$$R_2NH \xrightarrow[\text{or anhydride}]{RCOCl} R_2N{-}C(=O){-}R$$

$$R_2NH \xrightarrow{ArSO_2Cl} \text{sulfonamides} \quad ArSO_2NR_2$$

9.16 Reaction of Primary Amines with Nitrous Acid

Primary amines to produce a nitrosamine, where one "R" is a hydrogen.

$$\underset{(1°,\ 2°,\ \text{or } 3°)}{R_2NH} \xrightarrow[\text{"NO}^{+}\text{"}]{NaNO_2/H_2SO_4,\ 0\ °C} \underset{\text{nitrosamine}}{R_2N{-}NO}$$

A further reaction takes place under the acidic conditions when a primary amine is involved.

$$R\overset{+}{N}H_2{-}N{=}\ddot{O} \xrightarrow{H^+} RNH{-}N{=}\overset{+}{O}H \longrightarrow RNH{=}N{-}OH$$

$$\longrightarrow R{-}\overset{+}{N}{\equiv}N \longrightarrow \text{alkene, ROH, rearrangement, } \textit{etc.}$$

9.17 Reaction of Secondary Amines with Nitrous Acid

Secondary amines first form a protonated nitrosamine, which subsequently loses the proton to give the dialkyl nitrosamine.

$$R_2\overset{+}{N}(H){-}NO \longrightarrow R_2N{-}NO$$

9.18 Reaction of Tertiary Amines with Nitrous Acid

Tertiary amines form a quaternary nitrosamine salt. This subsequently breaks down to a dialkyl nitrosamine and an alcohol.

$$R_3\overset{+}{N}{-}NO \xrightarrow{H_2O} R_2N{-}NO + ROH$$

9.19 The Carbylamine Reaction

This reaction produces an isonitrile from a primary amine and a carbene produced from chloroform by the action of a strong base.

$$R{-}NH_2 + CHCl_3 \xrightarrow[\text{ethanol}]{KOH} R{-}\overset{+}{N}{\equiv}\overset{-}{C} \longleftrightarrow R{-}N{=}C{:}$$

$$CHCl_3 \xrightarrow{KOH} {:}CCl_2 \xrightarrow{R-\ddot{N}H_2} R{-}\overset{+}{N}H_2{-}CCl_2 \longrightarrow R{-}\overset{+}{N}{\equiv}\overset{-}{C}$$

dichlorocarbene

9.20 Oxidation of Tertiary Amines

$$R_3N \xrightarrow{H_2O_2} R_3\overset{+}{N}{-}\overset{-}{O}$$

powerfully basic amine oxide

9.21 Hofmann Elimination of Quarternary Ammonium Hydroxide

Amine oxides can be converted to hydroxylamines and an alkene by the action of heat. The mechanism through which this reaction proceeds is shown below. Effectively the amine oxide can act as a good leaving group.

$$R_3\overset{+}{N}-\overset{-}{O} \xrightarrow{250\,^\circ C} R_2NOH + \text{alkene}$$

$$H-C-C-\overset{+}{N}R_2-O^- \longrightarrow C=C + R_2NOH$$

$$R^1R^2R^3N \xrightarrow{R^4X} R^3-\overset{+}{N}(R^1)(R^2)-R^4 \xrightarrow{\text{heat}} R^1R^2R^3N + \text{alkene}$$

X = halogen

9.22 Reaction of Amines with Carbon Disulfide

Amines react with carbon disulfide to produce an isothiocyanate.

$$R-NH_2 \xrightarrow{CS_2} RNH-C(=S)S^- \;\; R\overset{+}{N}H_3 \xrightarrow[H_2O]{AgNO_3} RNH-C(=S)S^- \;\; Ag^+$$

$$\xrightarrow[H_2O]{\text{heat}} Ag_2S + H_2S + R-N=C=S$$

isothiocyanate

9.23 Relevant Questions

34(c), 48(c), 106(a), 126(a), 144(c), plus parts of functional group interconversion questions.

10 THE CARBONYL GROUP

(See Warren 1985.)

10.1 General Background to the Carbonyl Group

The chemistry of the carbonyl group is central to organic chemistry. In general the carbonyl group is considered to be C=O attached to at least one alkyl or aryl group R. Carbon dioxide and carbon monoxide are not included under this heading.

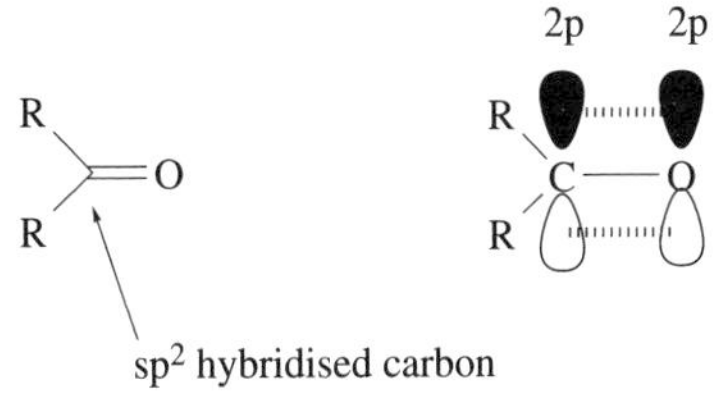

sp^2 hybridised carbon

The C=O bond is thermodynamically very stable. It has a large dipole moment (about 2.3 to 2.8 Debye units) due to the large difference in electronegativity between the carbon and the oxygen. The carbon is therefore electrophilic in nature.

The chemistry of the carbonyl group is hence dominated by nucleophilic attack.

The nucleophile (Nu = Nu: or Nu^-) approaches at an angle of 109° to the plane of the molecule. The nature of the groups R^1 and R^2 has an effect on the amount of positive charge there is on the carbonyl carbon and influences how the nucleophile can attack.

10.2 Principal Members of the Carbonyl Family

The principal members of the carbonyl family of compounds are named and shown below.

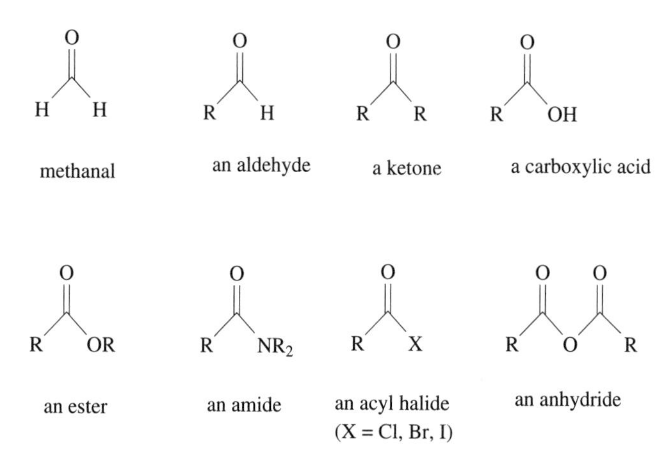

10.3 Factors Affecting Carbonyl Reactivity towards Nucleophiles

It is reasonable to suppose that the greater the degree of positive charge there is on the carbon the more susceptible the carbonyl group is to attack by a nucleophile. If we replace a hydrogen by an alkyl group then the alkyl group directs electron density towards the carbon of the carbonyl group reducing the reactivity, so, for example, methanal is more reactive than ethanal. A general series of reactivity can be built up.

RCOX, RCO–O–COR, HCOH, RCOH, RCOR, RCOOR, RCONR$_2$, RCOOH

decreasing susceptibility towards nucleophilic attack

X = a halogen

If the alkyl/aryl groups are large they may stop the nucleophile attacking the carbon of the C=O group.

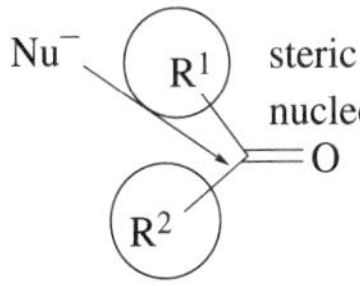
Nu⁻
R^1
R^2
O

steric hindrance may prevent nucleophilic attack

It is found that as R^1 and R^2 increase in size, nothing happens until a critical point is reached where R^1 and R^2 interfere with the "flight path" of the nucleophile. Large groups such as the *t*-butyl group definitely affect nucleophilic attack.

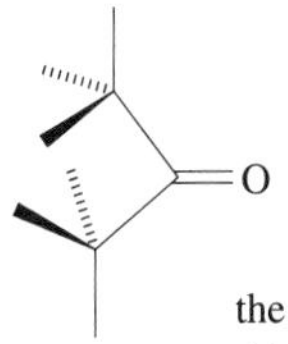
O

the *t*-butyl groups in this molecule would slow down nucleophilic attack considerably

10.4 Possibilities of Attack at a Carbonyl Group by a Nucleophile

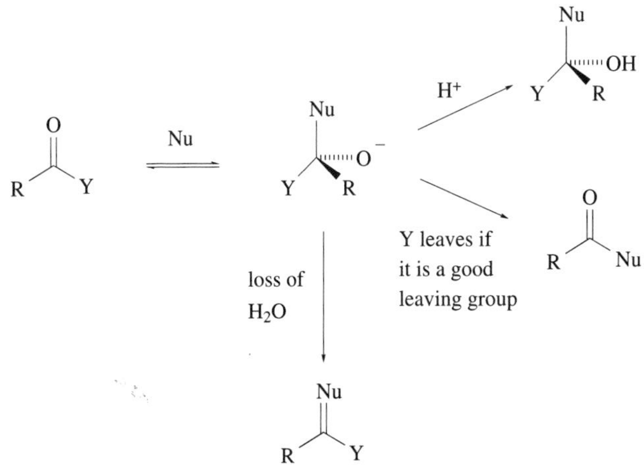
O
R
Y
Nu
Nu
O⁻
Y
R
H⁺
Nu
OH
Y
R
Y leaves if
it is a good
leaving group
O
R
Nu
loss of
H_2O
Nu
R
Y

10.5 Ester Groups

These are less reactive than ketones and aldehydes. The group is stabilised by the lone pair on the oxygen (—OR) feeding into the carbonyl group. This is partial delocalisation and stabilises the positive charge on the carbon.

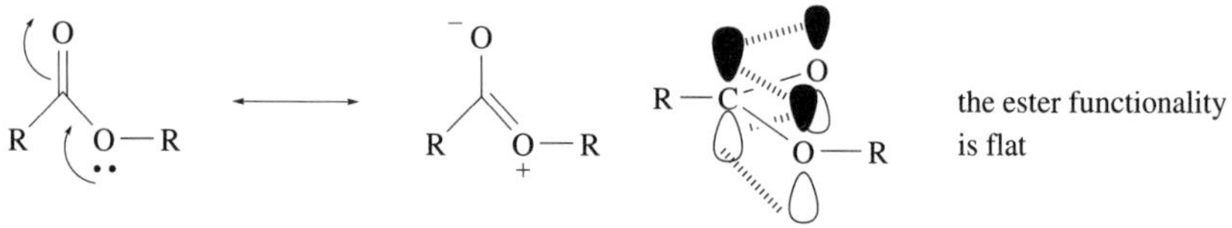

the ester functionality is flat

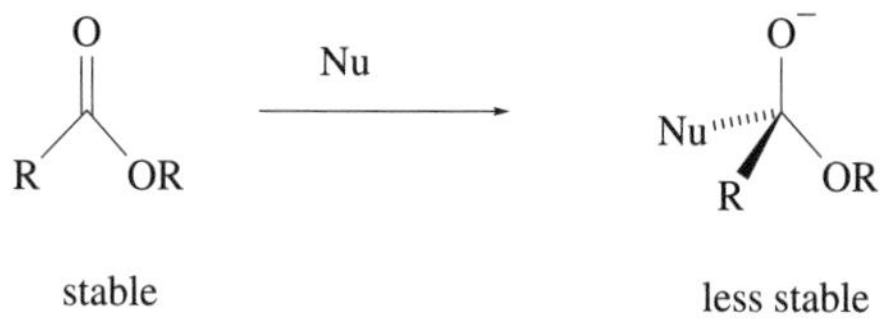

stable

less stable

This stabilisation is not present when the nucleophile has been added. There is therefore a decrease in reactivity towards nucleophilic attack.

10.6 Amide Groups

In the amide group there is donation of the lone pair on nitrogen into the carbonyl group. This is easier than in the ester case. Loss of this donation or delocalisation is destabilisation of the system.

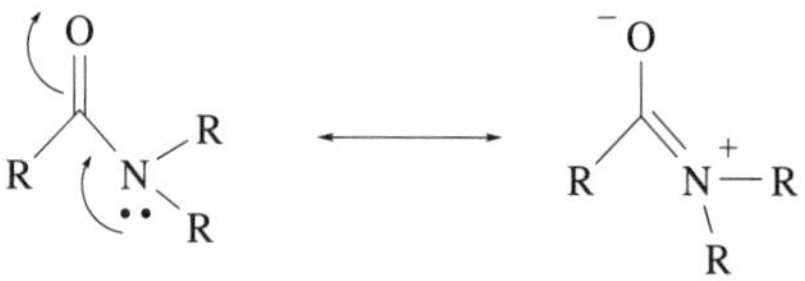

An amide is therefore one of the least reactive carbonyls. There can be primary, secondary and tertiary amides, depending on the extent of substitution at nitrogen.

10.7 Acid Chloride Groups

These are very susceptible to nucleophilic attack. There is poor overlap between the 3p orbital of chlorine and the 2p orbitals of carbon and oxygen, there is no groundstate stabilisation (delocalisation). Only the electronegativity contributes to the reactivity.

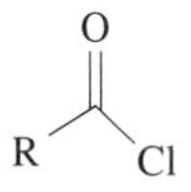

acyl chloride (sometimes called an acid chloride)

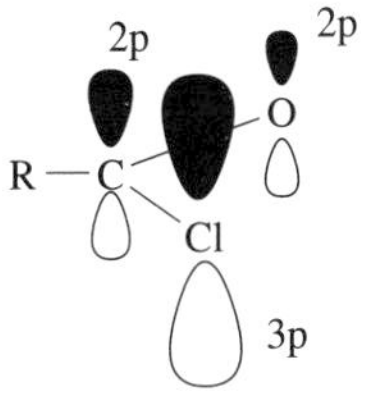

10.8 The Carbonyl Group and Acid Catalysis

The reaction of a nucleophile with a carbonyl compound is often catalysed by acid.

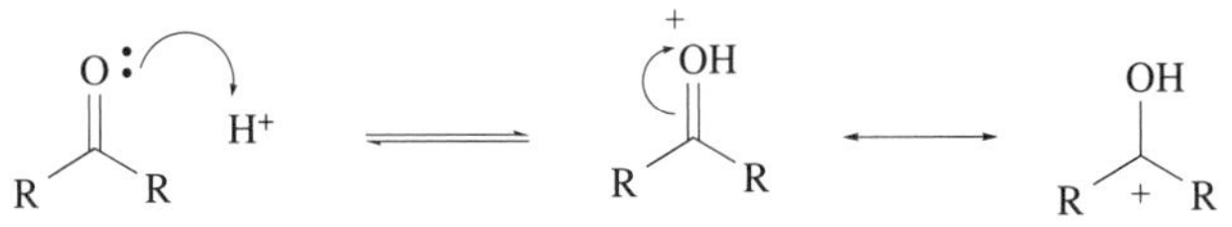

the increase of positive charge at carbon makes the carbonyl much more susceptible to nucleophilic attack

10.9 Preparation of Aldehydes and Ketones by Oxidation

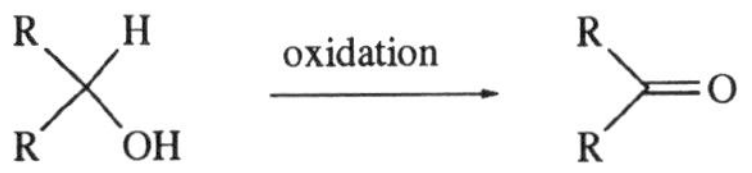

B is a base, B: or B^-

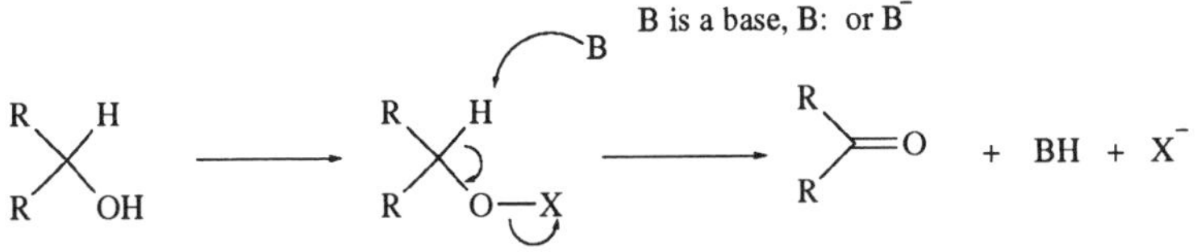

X is a good leaving group, forms stable X^-

If we want to prepare a ketone then this is a very easy reaction. Oxidants that can be used are numerous; CrO_3 in ethanoic acid or $K_2Cr_2O_7$ with H_2SO_4 (Jones reagent) are just two.

chromate ester

Another is bromine in water:

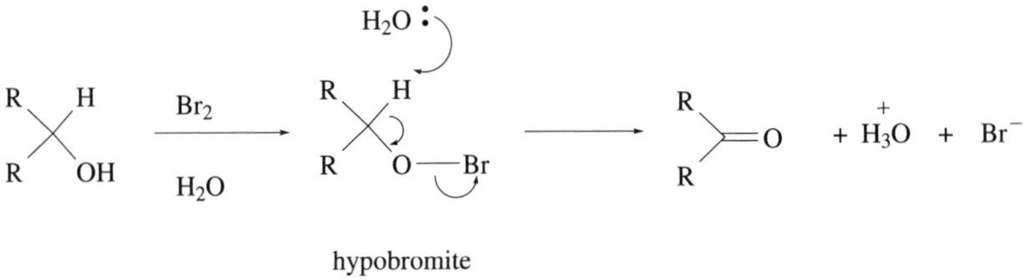

hypobromite

However if we want to form aldehydes problems occur. The reaction usually gives some acid as side product.

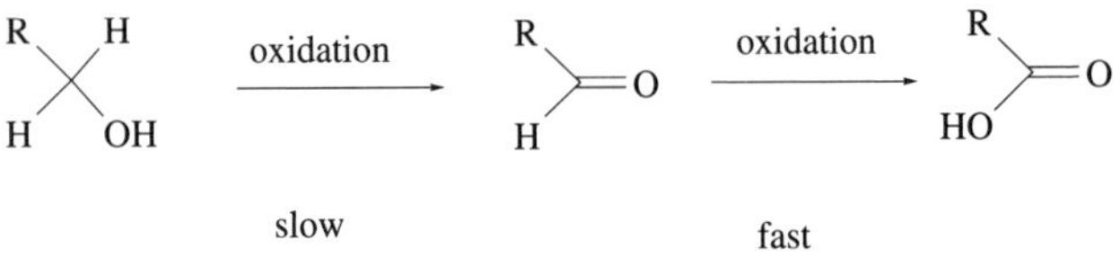

oxidation slow

oxidation fast

If we restrict ourselves to small molecules then they can be distilled out of the reaction mixture as formed, thus preventing further oxidation.

There are some methods available which proceed only to the aldehyde. One of these is Collin's reagent: CrO_3, and pyridine in dichloromethane. There is a variant of this reaction, pyridinium chlorochromate or Corey's reagent (and *t*-bu-$OCrO_3H$ can also be used for this type of oxidation).

Collin's reagent:

R, H, H, OH $\xrightarrow[\text{pyridine}]{CrO_3}$ R, H, H, O—Cr(=O)(=O)—N^+ (pyridine) ⟶ R, H, =O

Corey's reagent:

R, H, H, OH $\xrightarrow[\text{chlorchromate}]{\text{Pyridinium}}$ R, H, =O

Another good oxidation method is the Swern oxidation which involves dimethyl sulfoxide and some other reagent like trifluoroacetic anhydride or oxalyl chloride.

oxalyl chloride, DMSO (Me, Me, S=O) $\xrightarrow{ROH}$ Me, Me, $\overset{+}{S}$—O—C(=O)—C(=O)Cl; R, H, H, $\ddot{O}H$ $\xrightarrow{Et_3N}$

R, H, H, O—$\overset{+}{S}$(Me)(Me), :NEt_3 ⟶ R, H, =O + Me_2S

Allylic alcohols can be oxidised very easily to aldehydes or ketones using MnO_2 in petroleum ether at room temperature.

R, H, OH $\xrightarrow[\text{petroleum ether}]{MnO_2}$ R, =O

R = H or alkyl

The reaction mechanism involves a radical where the electron is delocalised throughout the system.

R OH ⟷ R OH

10.10 Preparation of Aldehydes and Ketones by Reduction of Carboxylic Acids and Derivatives

A similar problem to that found in oxidation arises for aldehydes when reduction is used to prepare them. However, there are viable methods.

R H H OH ← reduction, fast — R H =O ← reduction, slow — R HO =O

10.11 Rosenmund Reduction

R–C(=O)OH —($SOCl_2$, reflux)→ R–C(=O)Cl —(H_2/Pd on $BaSO_4$, N (quinoline))→ R–C(=O)H

10.12 Lithium Aluminium Hydride Reduction of Carboxylic Acids and Derivatives

R–C(=O)OH or R–C(=O)OR —($LiAlH_4$, ether)→ RCH_2OH

Lithium aluminium hydride will reduce an acid or ester all the way to an alcohol. Modification of lithium aluminium hydride by introducing *t*-butoxy groups in the place of three of the hydrogens in $LiAlH_4$

produces a reducing agent just capable of reduction of acids to aldehydes. Sodium borohydride, $NaBH_4$, can be used to reduce esters to alcohols but not acids.

$$LiAlH_4 \xrightarrow[\text{ether}]{ROH} LiAl(OR)H_3 \xrightarrow[\text{ether}]{ROH} LiAl(OR)_2H_2 \xrightarrow[\text{ether}]{ROH} LiAl(OR)_3H$$

decreasing reducing ability →

ROH = $(CH_3)_3C$–OH

$$RCOOH \xrightarrow{LiAl(OR)_3H} RCHO$$

The $LiAl(Ot\text{-}Bu)_3H$ is added slowly and is never in excess.

10.13 Aldehydes by Reduction of Nitriles Followed by Hydrolysis

$$R{-}C{\equiv}N \xrightarrow{\text{reduce}} \underset{\text{imine}}{RCH{=}NH} \xrightarrow{H_2O/H^+} RCHO$$

$$\underset{\text{imine}}{RCH{=}NH} \xrightarrow{\text{reduce}} \underset{\text{amine}}{RCH_2NH_2}$$

It is important not to reduce the nitrile all the way to an amine.

10.14 Stephen's Reduction of a Nitrile to give Aldehydes

$$R{-}C{\equiv}N \xrightarrow[HCl]{SnCl_2} RCH{=}N(H)SnCl_2 \xrightarrow{H_2O} RCHO$$

The conditions for this reaction limit what the alkyl/aryl group R is: it cannot be sensitive to the harsh conditions.

Reduction of nitriles using diisobutyllithium aluminium hydride is also possible.

$$R-C\equiv N \xrightarrow[-40°C]{i\text{-}Bu_2AlH} (H)(R)C=N-Al(i\text{-}Bu)_2 \xrightarrow{H_2O/H^+} R-C(=O)H$$

10.15 Ketones from Grignard Reagents with Nitriles, Amides or Acid Chlorides

Grignard reagents are organometallic compounds derived from an alkyl or aryl halide and magnesium.

$$RBr \xrightarrow{Mg} RMgBr$$

They will react with nitriles to ultimately form ketones.

$$R-C\equiv N \xrightarrow{R^1MgBr} (R^1)(R)C=N-MgBr \xrightarrow{H_2O/H^+} (R^1)(R)C=N-H$$

$$\xrightarrow{H_2O/H^+} R-C(=O)-R^1$$

They will also react with primary amides to form ketones.

$$R-C(=O)-NH_2 \xrightarrow{R^1MgBr} \left[R-C(OMgBr)(NH_2)-R^1\right] \xrightarrow{H_2O/H^+} R-C(OH)(NH_2)-R^1$$

$$\longrightarrow R-C(=O)-R^1$$

It is also possible to take an acid chloride with a Grignard reagent and form ketones.

$$\mathrm{RCOCl} \xrightarrow{R^1MgBr} \mathrm{RCOR^1}$$

R^1MgBr added to acyl chloride in THF ("reverse addition")

There are no equivalent reactions in general to form aldehydes. However, it is possible to use ethyl methanoate to form aldehydes.

$$\mathrm{HCO{-}OEt} \xrightarrow{R^1MgBr} \mathrm{HCOR^1}$$

ethylmethanoate 60% yield

10.16 Ketones from Calcium Salts of Acids

$$(RCO_2)_2Ca \xrightarrow{\text{heat}} \mathrm{RCOR} + CO_2 + CaO$$

If thorium is used instead of calcium then very high yields are obtained. It is difficult to produce cyclic ketones as the graph below shows.

for $(CH_2)_n$ C=O (ring)

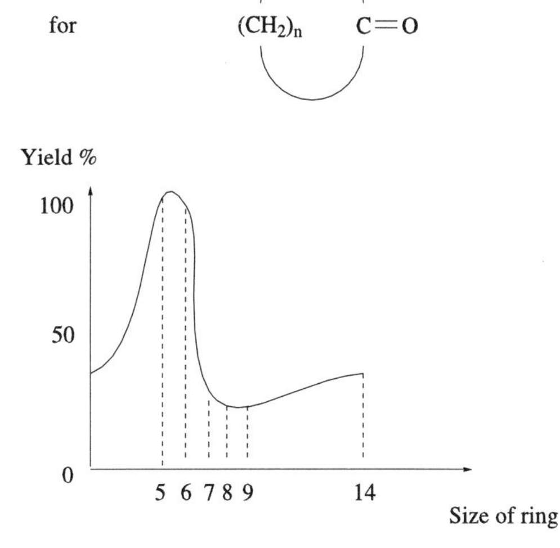

When $n = 4$ or 5 they are easy to make. From 7 upwards the difficulty increases. The use of thorium with diacids gives a method of making these larger rings.

10.17 Aldehydes and Ketones by Hydration of Alkynes

See 4.8 and 4.9.

10.18 Fission of Diols to Give Aldehydes and Ketones

The intermediates for these reactions are cyclic.

$$R_2C(OH)-C(OH)R_2 \xrightarrow{NaIO_4/H_2O} \left[\text{cyclic } R_2C(O)-C(O)R_2\, IO_3H\right] \longrightarrow 2\ R_2C{=}O$$

$$R_2C(OH)-C(OH)R_2 \xrightarrow{OsO_4} \left[\text{cyclic } R_2C(O)-C(O)R_2\, OsO_2\right] \longrightarrow 2\ R_2C{=}O$$

$$R_2C(OH)-C(OH)R_2 \xrightarrow{Pb(OAc)_4} \left[\text{cyclic } R_2C(O)-C(O)R_2\, Pb(OAc)_2\right] \longrightarrow 2\ R_2C{=}O$$

These types of reaction are often used when the diol is part of a ring or when one of the products is easy to remove from the reaction: for example, if one product is methanal (highly volatile).

10.19 Aldehydes and Ketones by Ozonolysis of Double Bonds

See 3.29.

10.20 Ketones from Diazomethane and Acid Chlorides

Halogenated ketones (see also 10.90 and 10.91) can be formed by the reaction of acid chlorides with diazomethane.

$$R{-}C(=O){-}Cl \xrightarrow[\text{excess}]{CH_2N_2} R{-}C(=O){-}CH_2Cl$$

10.21 Reactions of Aldehydes and Ketones: Nucleophilic Attack by Water

$$R_2C{=}O + H_2O \rightleftharpoons H_2\ddot{O}: \; R_2C{=}\overset{+}{O}H$$

aldehyde or ketone

$$R_2C(OH)(\overset{+}{O}H_2) \rightleftharpoons R_2C(OH)_2$$

gem(*inal*) diol or hydrate

The position of the equilibrium depends on the alkyl groups R attached. With ketones the equilibrium lies to the left. With certain aldehydes the equilibrium is almost exclusively to the right.

$$Cl_3C{-}CHO \underset{H_2SO_4}{\overset{H_2O}{\underset{\text{exothermic}}{\rightleftharpoons}}} Cl_3C{-}CH(OH)_2$$

chloral hydrate

10.22 Reactions of Aldehydes and Ketones: Nucleophilic Attack by Alcohols

Alcohols react with aldehydes and ketones if there is an acid or Lewis acid catalyst around.

An alcohol can attack once to give a hemiacetal:

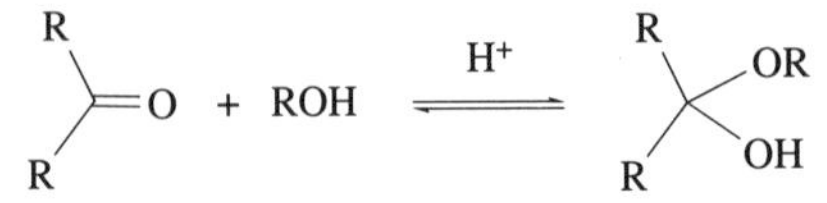

$$R_2C{=}O + ROH \underset{}{\overset{H^+}{\rightleftharpoons}} R_2C(OR)(OH)$$

aldehyde or ketone — hemiacetal

and then another alcohol molecule can attack the hemiacetal to give an acetal.

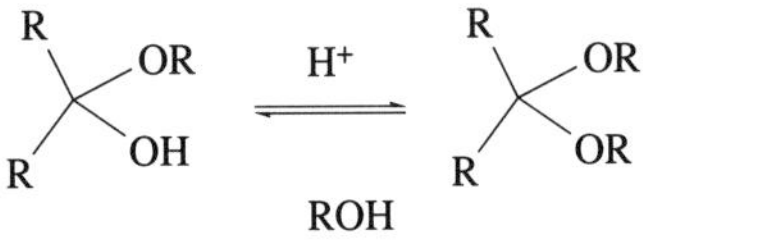

$$R_2C(OR)(OH) \underset{ROH}{\overset{H^+}{\rightleftharpoons}} R_2C(OR)(OR)$$

hemiacetal — acetal (or ketal)

The mechanism for this reaction is:

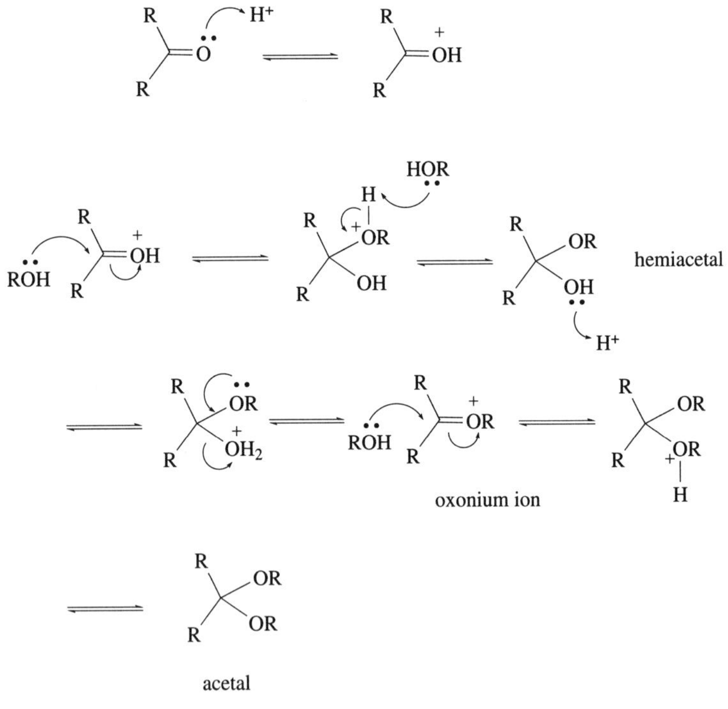

$$R_2C{=}\ddot{O} + H^+ \rightleftharpoons R_2C{=}\overset{+}{O}H$$

$$R\ddot{O}H + R_2C{=}\overset{+}{O}H \rightleftharpoons R_2C(\overset{+}{O}(H)R)(OH) \xrightarrow{HOR} R_2C(OR)(\ddot{O}H) \text{ hemiacetal} \xrightarrow{H^+}$$

$$\rightleftharpoons R_2C(\ddot{O}R)(\overset{+}{O}H_2) \rightleftharpoons R\ddot{O}H + R_2C{=}\overset{+}{O}R \rightleftharpoons R_2C(OR)(\overset{+}{O}(H)R)$$

oxonium ion

$$\rightleftharpoons R_2C(OR)(OR)$$

acetal

Acetals can be used to protect the ketone and aldehyde functionalities whilst other reactions that would affect the groups take place. The acetal can be converted back to the carbonyl compound at a later stage.

Acetals are stable under neutral and alkaline conditions. An alcohol often used for protection is 1,2-ethanediol.

HO OH

remove H_2O formed (Dean and Stark)

10.23 Reactions of Aldehydes and Ketones: Nucleophilic Attack by Thiols

Thiols react in a similar manner to alcohols with ketones and aldehydes. A commonly used thiol for protection is ethanedithiol. It is harder to convert a dithioacetal back to the carbonyl compound than an acetal. This is because of the orbitals involved in the hydrolysis. The hydrolysis of acetals or thioacetals proceeds through the oxonium or sulfonium ion. This requires an overlap of orbitals, as illustrated below. The overlap is easier to achieve for C–O than for C–S because the orbitals are the same size. Mercury salts are often used to aid the hydrolysis.

HS SH

H^+

$HgCl_2/H_2O$

2p 2p

positive charge build-up

good orbital overlap

leaving alcohol

formed easily compared to the analogous thionium ion

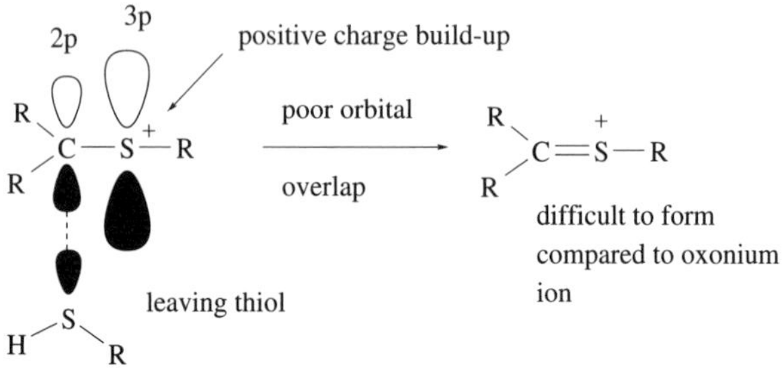

2p 3p positive charge build-up

poor orbital overlap

difficult to form compared to oxonium ion

leaving thiol

10.24 Desulfurisation of Dithioacetals

Reagents such as Raney nickel can be employed to convert the dithioacetal to a hydrocarbon.

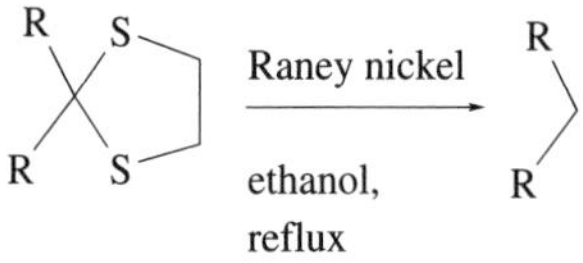

Raney nickel

ethanol, reflux

10.25 Reactions of Aldehydes and Ketones: Addition of Tertiary Amines

If a tertiary amine is added to an aldehyde or ketone an equilibrium is set up. No new compound can be formed.

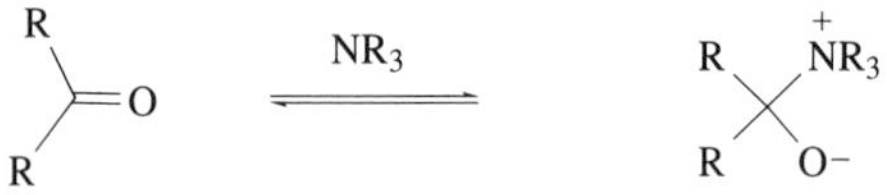

NR_3

10.26 Reactions of Aldehydes and Ketones: Addition of Ammonia

If ammonia is added to an aldehyde or ketone then trimers are produced.

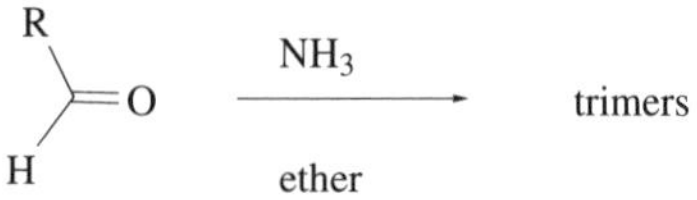

NH_3

ether

trimers

Methanal produces hexamethylenetetramine with ammonia.

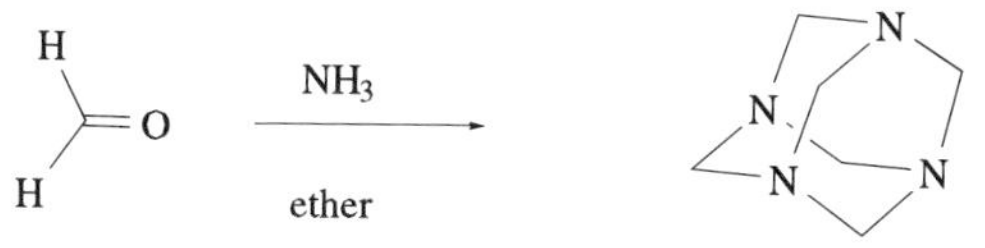

$$H_2C{=}O \xrightarrow[\text{ether}]{NH_3} \text{hexamethylenetetramine}$$

hexamethylenetetramine

However, if primary or secondary amines are reacted with aldehydes or ketones then two new types of organic molecule can be synthesised.

10.27 Reactions of Aldehydes and Ketones: Addition of Primary Amines

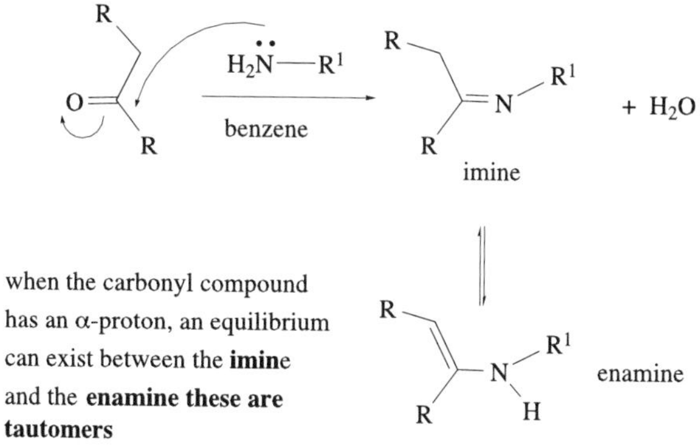

$$R_2C{=}O \xrightarrow[\text{benzene}]{H_2N{-}R^1} R_2C{=}N{-}R^1 + H_2O$$

imine

enamine

when the carbonyl compound has an α-proton, an equilibrium can exist between the **imine** and the **enamine these are tautomers**

There is still a problem with amines that have a small group R^1 attached. The imines produced tend to trimerise.

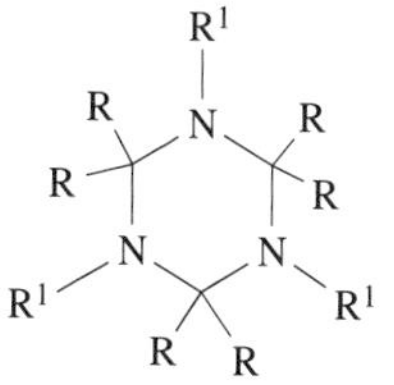

trimers if R^1 is small

The water produced in the reaction must be removed or the carbonyl compound and amine are re-formed. This is carried out using a Dean and Stark apparatus.

Reduction of the imine as it is produced gives rise to secondary amines, cyanoborohydride can also be used for this.

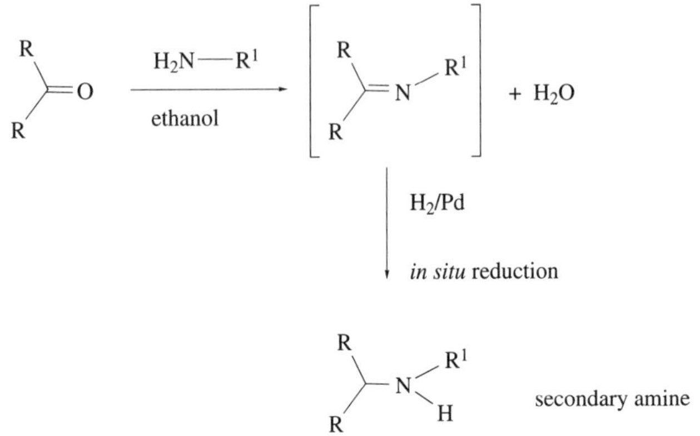

H_2N—R^1
ethanol
+ H_2O
H_2/Pd
in situ reduction
secondary amine

10.28 Reactions of Aldehydes and Ketones: Addition of Secondary Amines

Secondary amines react with aldehydes and ketones to give enamines. There must, however, be an α-proton on the aldehyde or ketone.

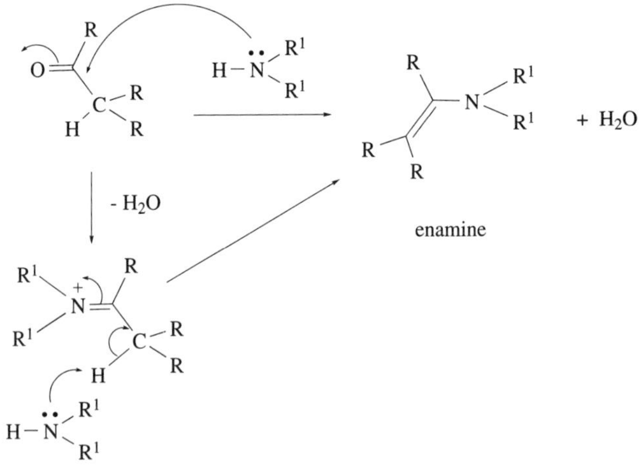

- H_2O
+ H_2O
enamine

Enamines function as nucleophiles:

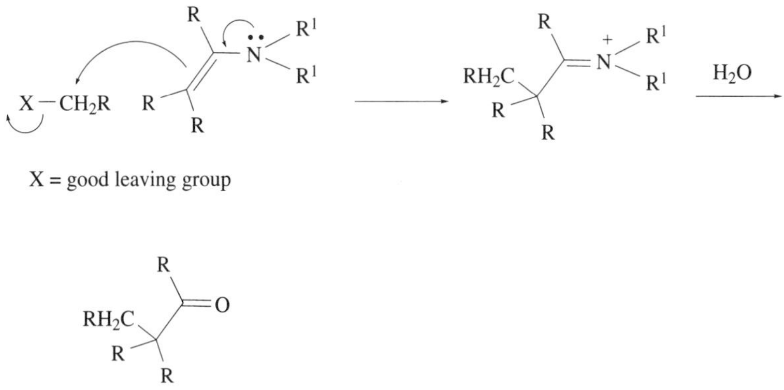

X = good leaving group

This is a method of alkylating an α-carbon.

10.29 Reactions of Aldehydes and Ketones: Addition of Hydroxylamine

Carbonyl compounds react with hydroxylamine to form an oxime.

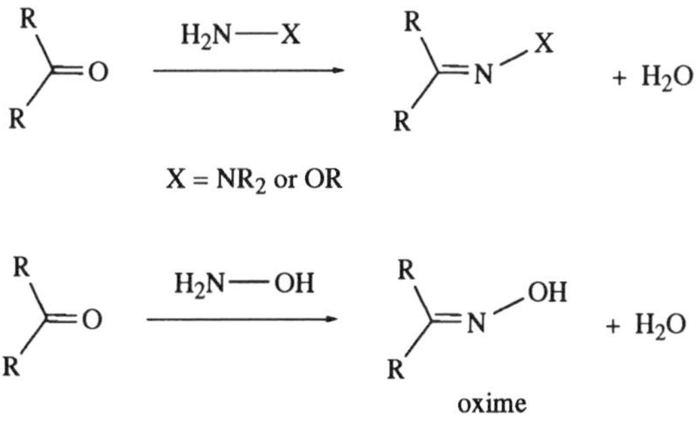

$X = NR_2$ or OR

oxime

There is a degree of conjugation in the molecule. This stabilises the molecule and pushes the equilibrium to the right.

The rate of formation of oximes is dependent on the pH. This is an acid-catalysed reaction.

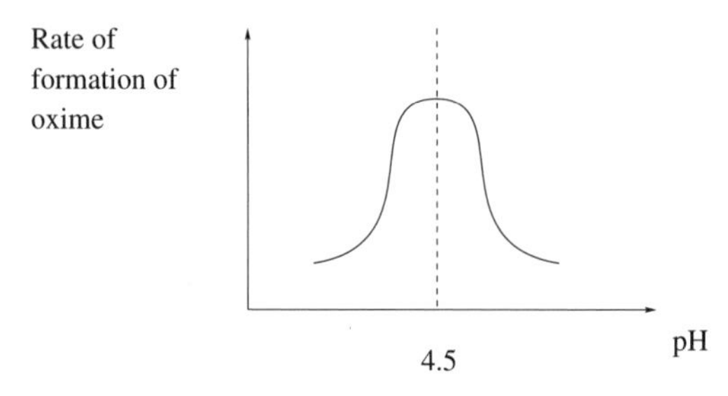

The mechanism of the reaction is as follows:

$$R_2C{=}O \xrightleftharpoons{H^+} R_2C{=}\overset{+}{O}H$$

$$HO{-}\ddot{N}H_2 + R_2C{=}\overset{+}{O}H \longrightarrow R_2C(OH){-}\overset{+}{N}H_2(OH) \quad + H_2O \longrightarrow \text{oxime}$$

but another equilibrium exists because the hydroxylamine is a base.

$$HO{-}NH_2 \xrightleftharpoons{H^+} HO{-}\overset{+}{N}H_3 \quad \text{not a nucleophile}$$

Therefore at low pH there is a lack of nucleophilic species around, so the rate decreases as the pH decreases. If there are not enough protons around then the amount of $C{=}OH^+$ around is small and so if the solution becomes more alkaline (higher pH) the rate also decreases.

$$H_3\overset{+}{N}OH\,\overset{-}{Cl} / \overset{+}{Na}\,\overset{-}{O}_2CCH_3 \quad \text{gives the correct pH}$$

Hydroxylamines can exist as geometrical isomers.

$$R^1R^2C{=}\ddot{N}{-}OH \qquad R^1R^2C{=}\ddot{N}{-}OH$$

10.30 The Beckmann Rearrangement

This rearrangement occurs when you take an oxime, derived from a ketone, and treat it with an acid, for example concentrated H_2SO_4, and pour it into water; an amide is formed.

$$R^1R^2C{=}N{-}OH \xrightarrow[H_2SO_4]{conc.} R^1{-}C(=O){-}NHR^2$$

derived from a ketone — amide

If the oxime is not derived from a ketone but an aldehyde, then a nitrile is formed.

$$R^1HC{=}N{-}OH \xrightarrow[H_2SO_4]{conc.} R^1CN$$

derived from an aldehyde — nitrile

The mechanism for formation of the amide is as follows:

$$R^1R^2C{=}N{-}OH \overset{H^+}{\rightleftharpoons} R^1R^2C{=}N{-}\overset{+}{O}H_2 \longrightarrow \underset{A}{R^2{-}C{\equiv}N^+{-}R^1} \longleftrightarrow \underset{B}{R^2{-}C^+{=}N{-}R^1} \quad :OH_2$$

$$\longrightarrow R^2(\overset{+}{O}H_2)C{=}N{-}R^1 \rightleftharpoons R^2(OH)C{=}N{-}R^1 \longleftrightarrow R^2{-}C(=O){-}NH{-}R^1$$

The group which migrates to the nitrogen is usually *anti* (*trans*) to the hydroxyl.

10.31 The Ritter Reaction

The cations A and B can be generated *via* the Ritter reaction. The Ritter reaction is primarily a reaction between a tertiary carbocation and a nitrile. For the reaction to proceed a source of tertiary carbocations is needed. The carbocation is usually derived from an alcohol or alkene.

The mechanism of the reaction is as follows:

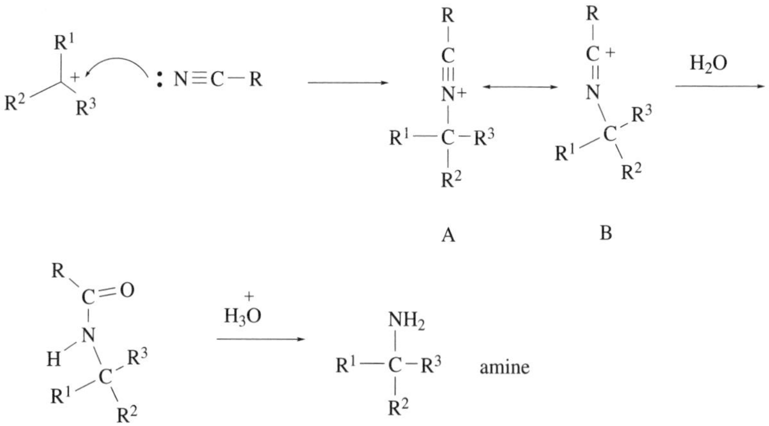

A

B

amine

Hydrolysis of the amide that is formed by refluxing with acid gives the tertiary amine.

10.32 Reaction of Ketones and Aldehydes with Hydrazine and Substituted Hydrazine

Aldehydes and ketones react with excess hydrazine to form hydrazones.

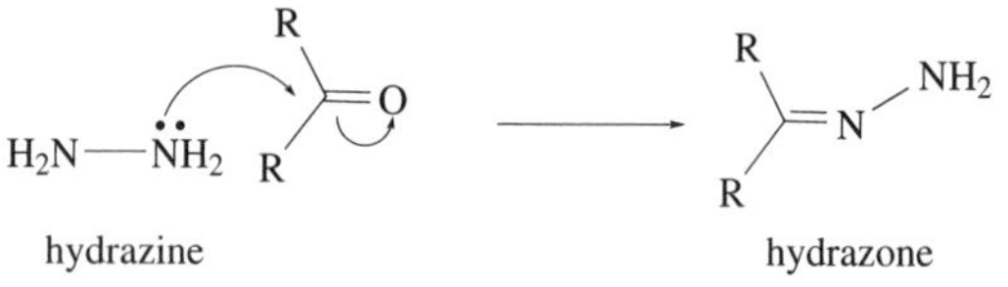

hydrazine

hydrazone

If there is not an excess of hydrazine, then an azine is produced.

azine

Hydrazones can be oxidised to form diazo compounds. This is an easy reaction if one group R is hydrogen.

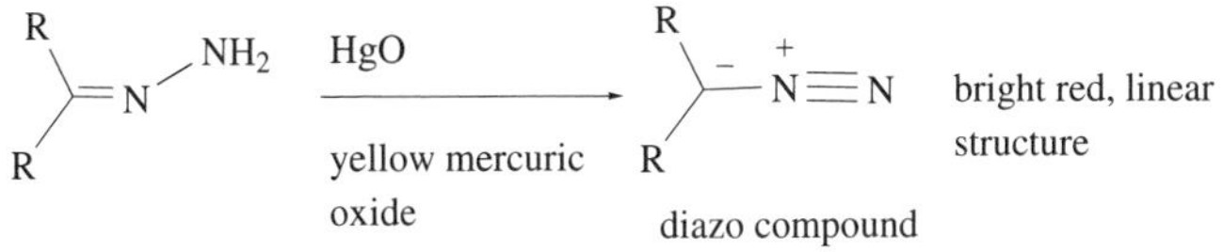

HgO

yellow mercuric oxide

diazo compound

bright red, linear structure

Substituted hydrazines such as semicarbazide and 2,4-dinitrophenylhydrazine (DNPH) react to give hydrazone derivatives.

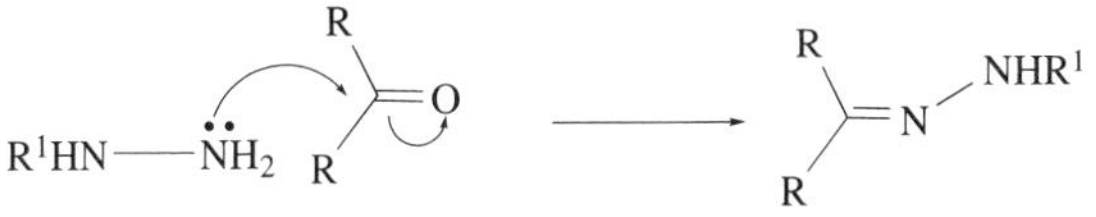

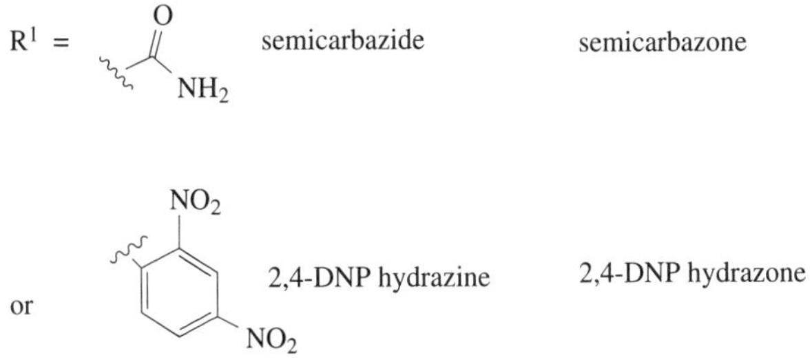

R^1 = semicarbazide semicarbazone

or 2,4-DNP hydrazine 2,4-DNP hydrazone

10.33 Formation of Cyanohydrins

A ketone or aldehyde will react with hydrogen cyanide to form a cyanohydrin.

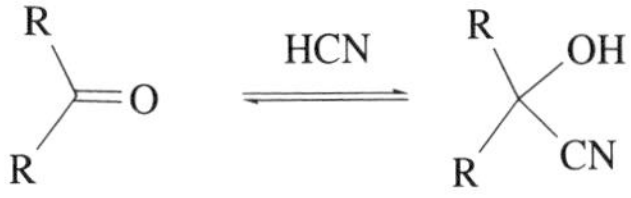

HCN

There is a need for a proton source; HCN on its own is not good at providing this. Therefore the reagents used are KCN (or NaCN) with a mineral acid.

$$R_2C{=}\ddot{O} \xrightleftharpoons{H^+} R_2C{=}\overset{+}{O}H \;(+\;^{-}CN) \rightleftharpoons R_2C(OH)CN$$

With aldehydes the equilibrium is usually well to the right-hand side; there is a high yield of cyanohydrin. With ketones the position of the equilibrium is very dependent on the groups attached to the carbonyl functionality.

Cyclohexanone ⟶ 100%

$Me_2C{=}O$ ⟶ 84%

$(CH_2)_{10}$ C=O (cyclic) ⟶ no reaction

Hydrolysis, by refluxing with a mineral acid, of a cyanohydrin results in an α-hydroxycarboxylic acid being formed.

$$R_2C(OH)CN \xrightarrow[\text{acid reflux}]{\text{mineral}} R_2C(OH)CO_2H$$

10.34 The Strecker Reaction

This is a reaction which allows the preparation of α-amino acids.

$$R_2C{=}O \xrightarrow{NH_4CN} R_2C(NH_2)CN \xrightarrow[\text{reflux}]{H^+} R_2C(NH_2)CO_2H$$

aldehyde or ketone; α-amino acid

10.35 The Baeyer–Villiger Oxidation

Ketones can be oxidised, using a peroxy acid, to an ester. The reaction is a rearrangement where one of the alkyl/aryl groups of the ketone migrates to form the ester.

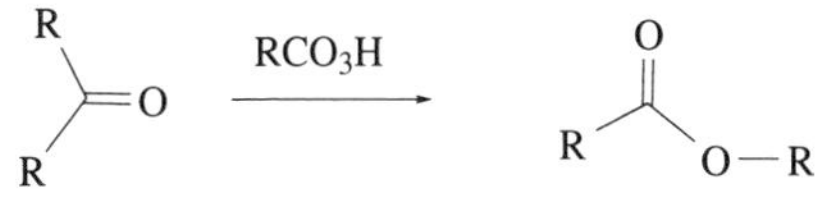

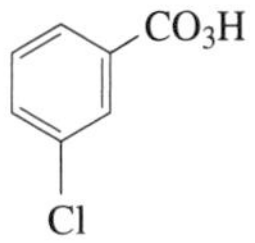

$$R_2C{=}O \xrightarrow{RCO_3H} RC(=O){-}O{-}R$$

MCPBA (*meta*-chloroperbenzoic acid)

10.36 Mechanism of the Baeyer–Villiger Oxidation

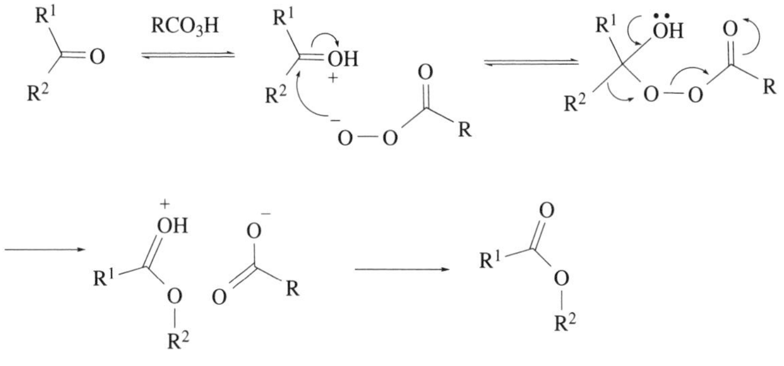

There is a preference order for the migration of alkyl groups:

tertiary > secondary > primary > Me.

10.37 Reaction of Aldehydes, Ketones and Esters with Grignard Reagents

An aldehyde, ketone or ester will react with a Grignard reagent to produce an alcohol.

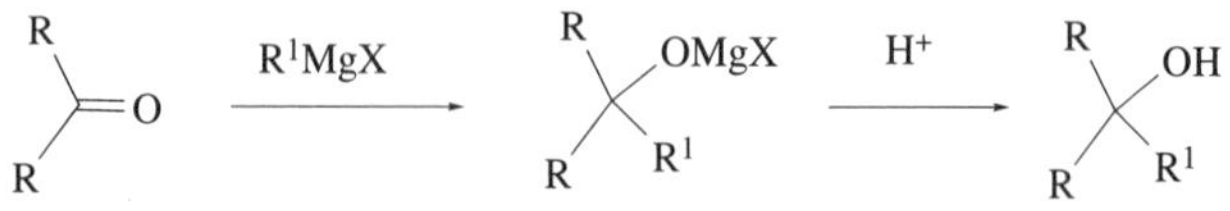

R(R)C=O —R^1MgX→ R(R)C(OMgX)R^1 —H^+→ R(R)C(OH)R^1

aldehyde, ketone or ester

X = halogen

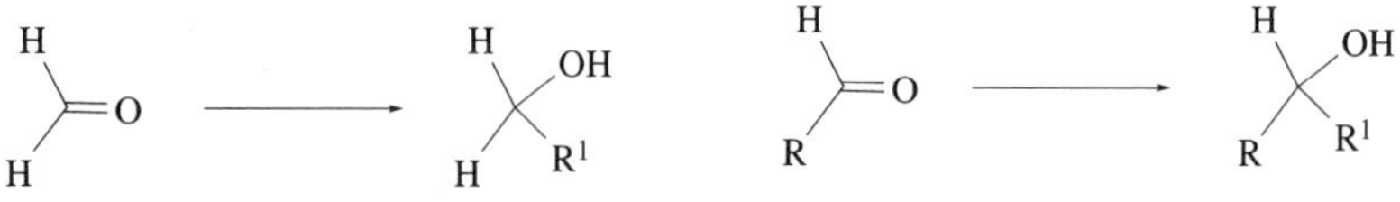

H(H)C=O ⟶ H(H)C(OH)R^1 H(R)C=O ⟶ H(R)C(OH)R^1

methanal primary alcohol, difficult aldehyde secondary alcohol

R(R)C=O ⟶ R(R)C(OH)R^1

ketone tertiary alcohol

R(RO)C=O ⟶ R(R)C=O ⟶ R(R)C(OH)R^1

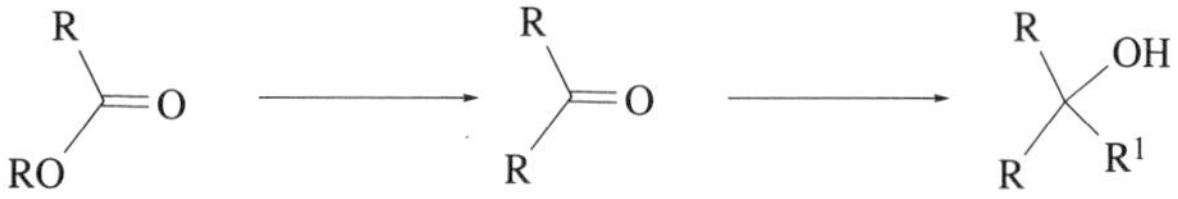

ester ketone tertiary alcohol

10.38 Reduction of Aldehydes and Ketones *via* the Dithioacetal to a Methylene Group

See 10.24.

10.39 The Clemenson Reduction of Aldehydes and Ketones to a Methylene Group

If a ketone or aldehyde is heated at reflux temperature with amalgamated zinc (zinc and mercury) and concentrated hydrochloric acid, it will be reduced to the corresponding hydrocarbon. This is only applicable to molecules stable under these conditions.

amal. Zn, aq. HCl

10.40 The Wolff–Kishner Reduction of Aldehydes and Ketones to a Methylene Group

These are strong basic conditions. The reaction goes through the mechanism below.

H_2NNH2; $NaOCH_2CH3$, 180 °C; NH_2; + N_2; H^+

The carbanion picks up a proton to give the product.

10.41 Pinacol by the Reduction of Ketones

Diols can be formed from ketones in the presence of magnesium activated by mercury chloride.

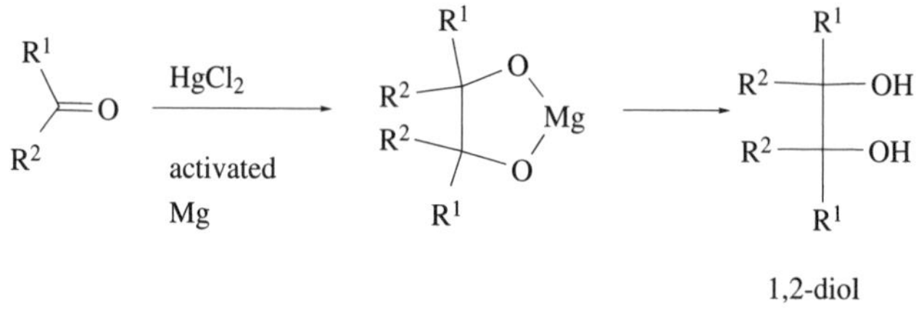

$HgCl_2$

activated Mg

1,2-diol

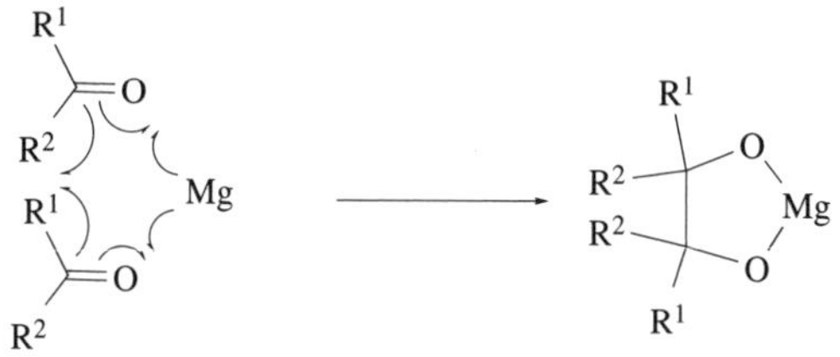

radical mechanism

The reaction of propanone (acetone) results in a compound whose trivial name is pinacol.

10.42 The Pinacol–Pinacolone Rearrangement

In the presence of acid certain diols will rearrange to a ketone or aldehyde. This is known as the pinacol–pinacolone rearrangement or, more technically, a 1,2-alkyl or 1,2-aryl shift.

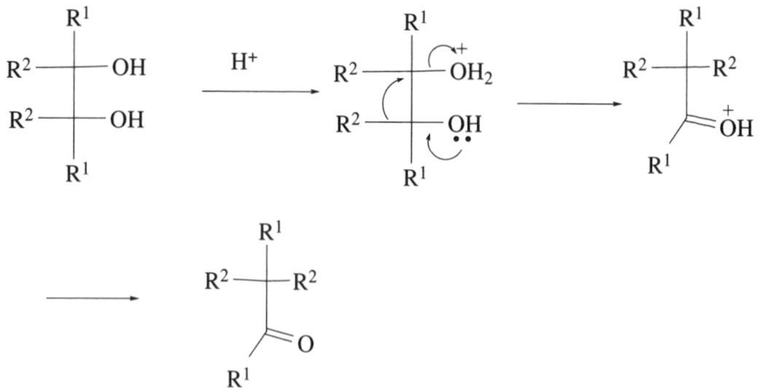

H^+

10.43 Reactions of Aldehydes and Ketones *via* Enolisation

An aldehyde or ketone can exist in two tautomeric forms, the enol and the carbonyl compound.

R—CH₂(R)C=O ⇌ R–CH=C(R)–OH

aldehyde or ketone — enol

The **enol** form is promoted by the **addition of acid**. The position of this equilibrium lies well to the left for simple aldehydes or ketones: there is about 6×10^{-7}% of enol in acetone and 1×10^{-3}% in acetaldehyde. Larger amounts of enol are present in aldehydes and ketones where conjugation and hydrogen bonding between a carbonyl and the hydrogen of the hydroxyl group can occur: for example, $CH_3COCH_2COCH_3$ has 76% enol content and $PhCOCH_2COCH_3$ has 89%. The position of the equilibrium is also very dependent on solvent, concentration and temperature.

conjugate base

H⁺

slow step

In the presence of a **base**, the carbonyl compound may be converted to the **enolate** by deprotonation.

base

enolate

Note that the hydrogen (or proton) adjacent to the carbonyl, or other stabilising group, which can be removed by base or undergo keto–enol tautomerism, is often called an active hydrogen, the α-hydrogen or acidic.

10.44 Halogenation of Aldehydes and Ketones

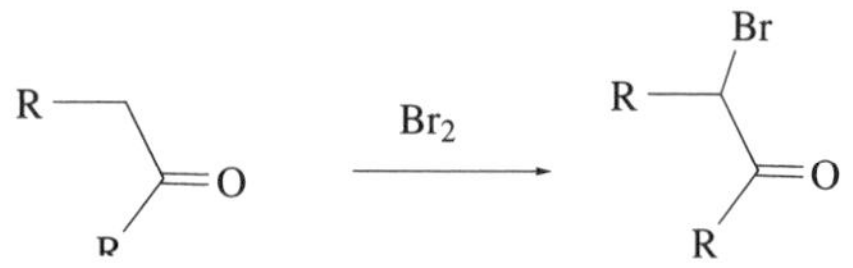

The reaction is catalysed by acid or base and the rate of reaction is proportional to [carbonyl][HO^-] (or [H^+]). The concentration of halogen is not involved in the rate equation, or in other words the halogen does not take part in the RLS.

10.45 Choice of Proton for Deprotonation in Unsymmetrical Ketones

If we have an unsymmetrical ketone, which proton (1 or 2) will be removed by the base?

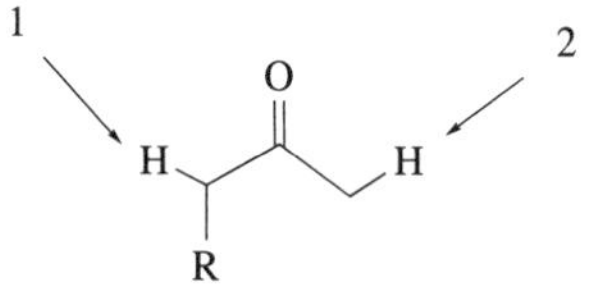

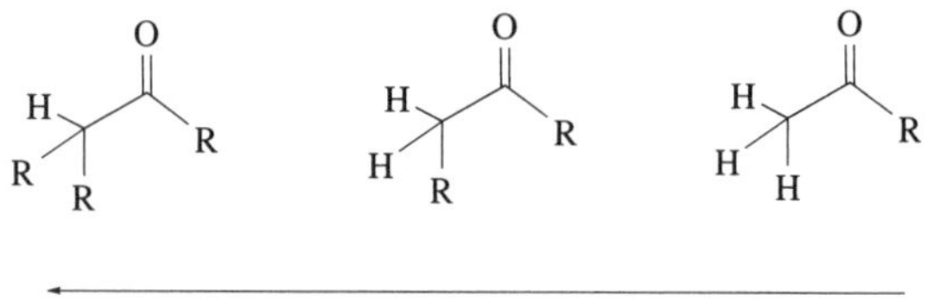

increasing ease of removal of proton by a base when R is an alkyl, aryl or stabilising group

A disadvantage with base-catalysed enolisation is that it increases the acidity of the two hydrogens on the same carbon as the bromine is introduced to. The base can now remove another proton and form the dibromo adduct, and then another to produce the tribromo compound.

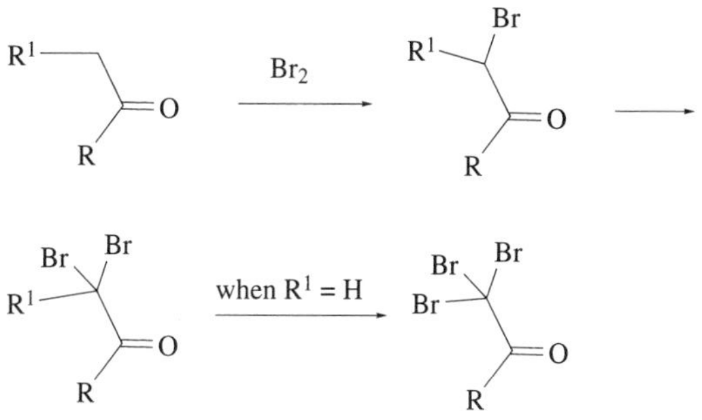

Br_2

when $R^1 = H$

This is the basis of the iodoform reaction (which is an analytical test for the presence of a methyl next to a carbonyl group).

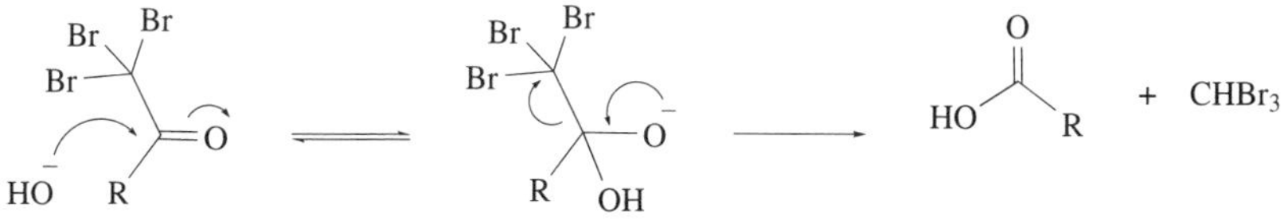

+ $CHBr_3$

The acid-catalysed halogenation reaction proceeds as follows:

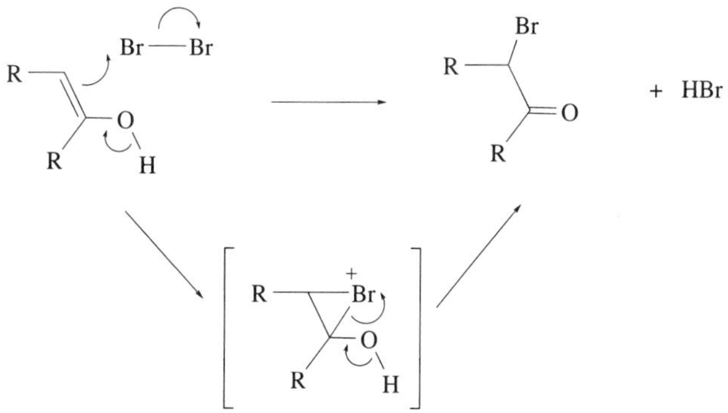

+ HBr

In the acid-catalysed halogenation there is no increase in rate as one proton is replaced by the halogen; therefore the reaction can be stopped after the first halogen is inserted.

10.46 The Aldol Condensation Reaction

The aldol condensation can be promoted by both acid and basic conditions. The base-catalysed reaction proceeds as shown below.

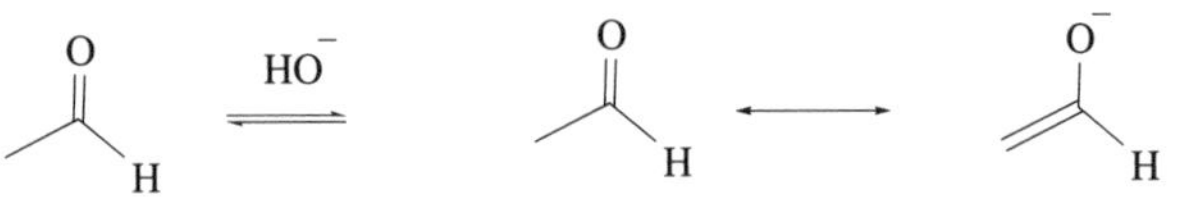

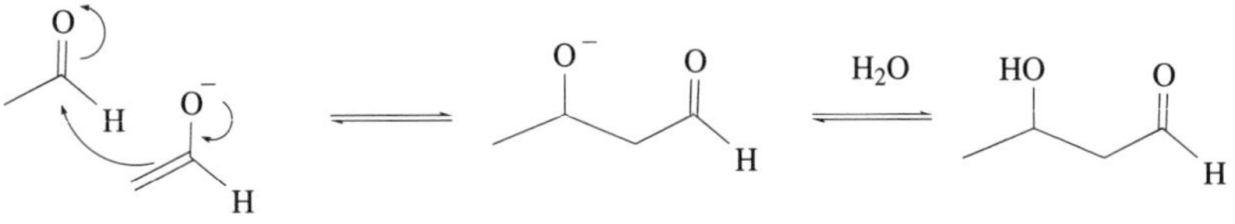

β-hydroxyaldehyde

3-hydroxybutanal

The base-catalysed reaction results in a *β*-hydroxyaldehyde or or *β*-hydroxyketone (if a betone was used to form the enolate) being formed. If these compounds are allowed to come into contact with acid then an *α*,*β*-unsaturated carbonyl compound is produced.

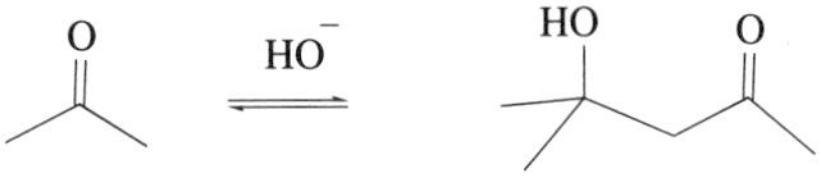

4-hydroxy-4-methyl-2-one

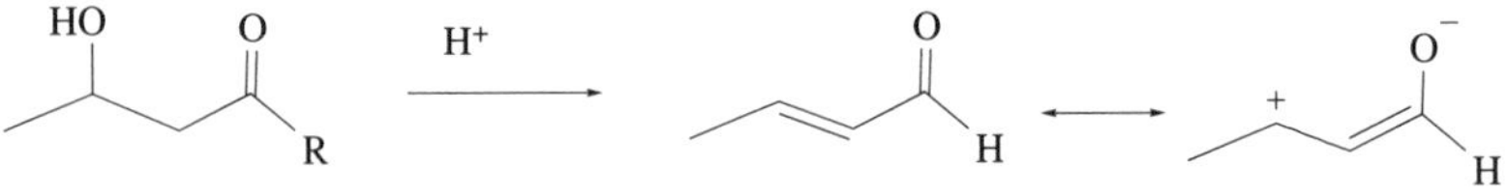

β-hydroxyaldehyde or ketone

α,β-unsaturated carbonyl

α,*β*-Unsaturated carbonyl compounds are stable and can be considered to exist in two resonance forms.

The acid-catalysed reaction proceeds by the mechanism shown below.

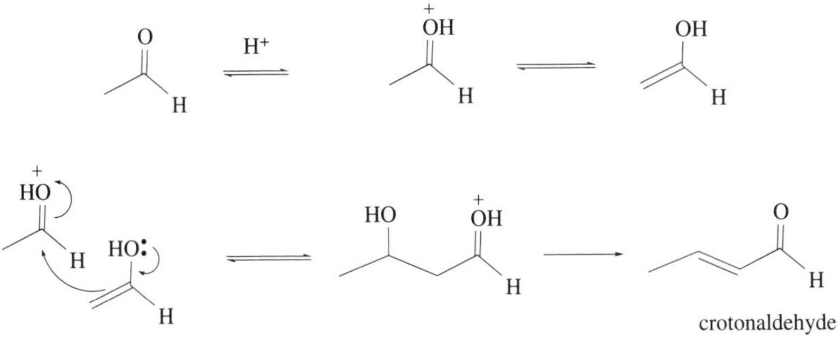

crotonaldehyde

10.47 Mixed or Crossed Aldol Reactions

If we take two different carbonyl compounds such as propanal and butanal we have the possibility of forming four products. This is not useful and it is hard to separate the products. This type of reaction is not attempted unless one of the reactants cannot react with itself.

This is achieved when one reactant has no α-protons

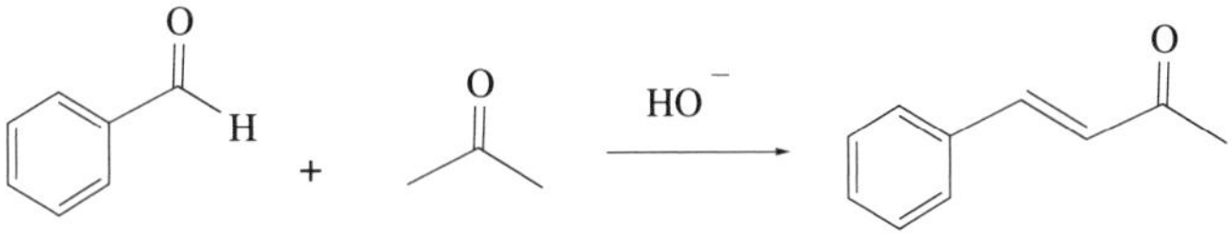

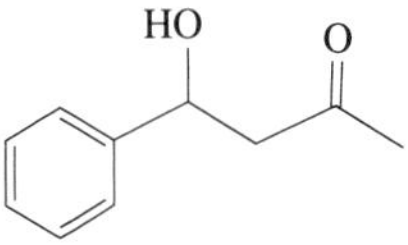

this intermediate cannot be isolated

Benzaldehyde has no α-protons and therefore cannot form an enol/enolate. Under vigorous basic conditions benzaldehyde or other carbonyls with no α-protons will react with themselves. This is known as the Cannizaro reaction

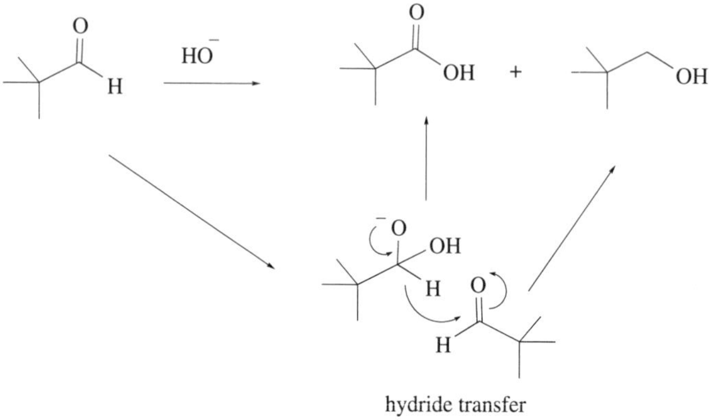

hydride transfer

10.48 The Claisen and Dieckmann Condensations

The Claisen condensation is when esters containing an α-hydrogen are treated with a strong base such as sodium methoxide; following this a condensation occurs to give a β-keto ester. If two different esters are used, each having an α-hydrogen, then there is a possibility of four products. Often one of the esters used in this reaction has no α-hydrogen, reducing the number of products.

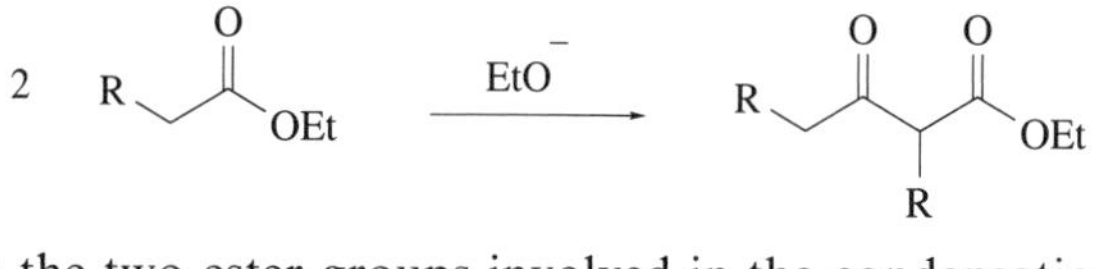

When the two ester groups involved in the condensation are in the same molecule, the product is a cyclic β-keto ester. This is known as the Dieckmann condensation and it works best for the formation of five-, six- and seven-membered rings.

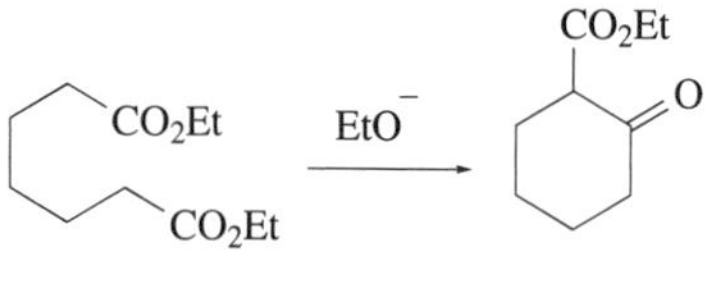

10.49 The Michael Addition Reaction

The Michael addition reaction involves conjugate addition of a nucleophile to an α,β-unsaturated carbonyl compound. However, other electron-withdrawing groups conjugated to the double bond will facilitate the reaction.

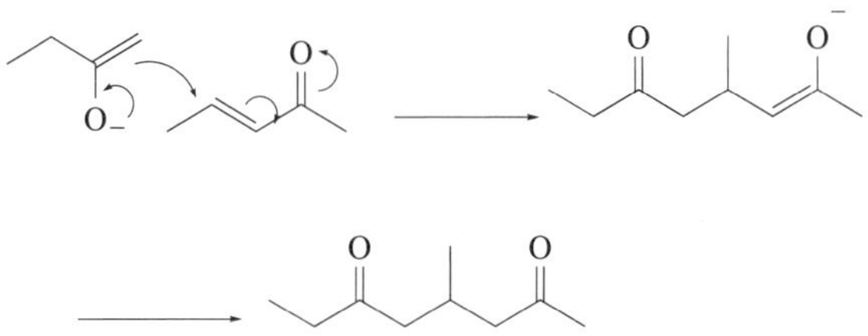

10.50 The Mannich Reaction

In the Mannich reaction, formaldehyde (or sometimes another aldehyde) is condensed with ammonia in the form of its salt, and a compound containing an active hydrogen. Alkyl amines (but usually not aryl amines) can be used in place of ammonia.

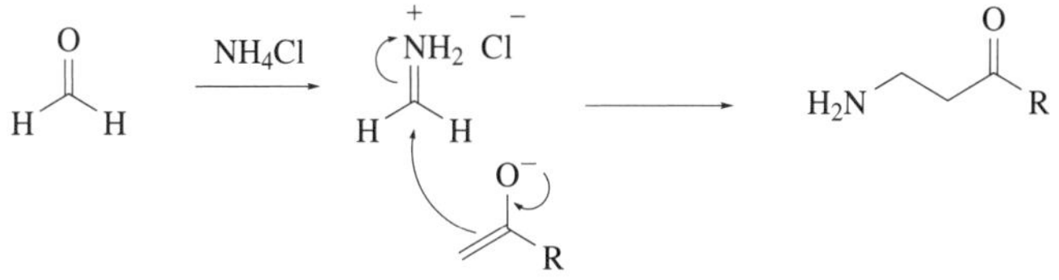

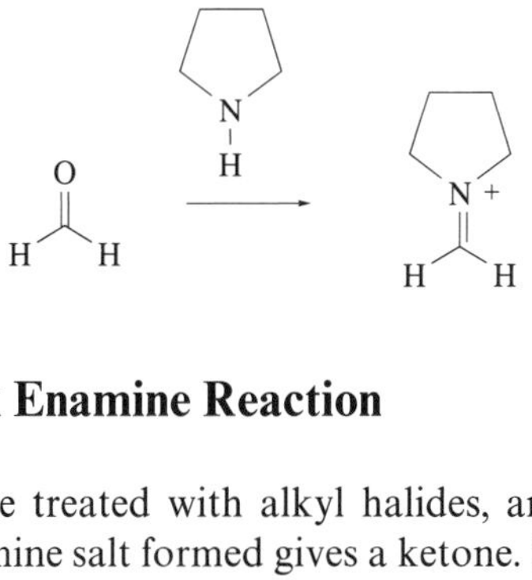

10.51 The Stork Enamine Reaction

When enamines are treated with alkyl halides, an alkylation occurs. Hydrolysis of the imine salt formed gives a ketone. Since the enamine is formed from a ketone the overall result is alkylation of the ketone at the α-position. This is an alternative to alkylation of ketones *via* the enol or enolate.

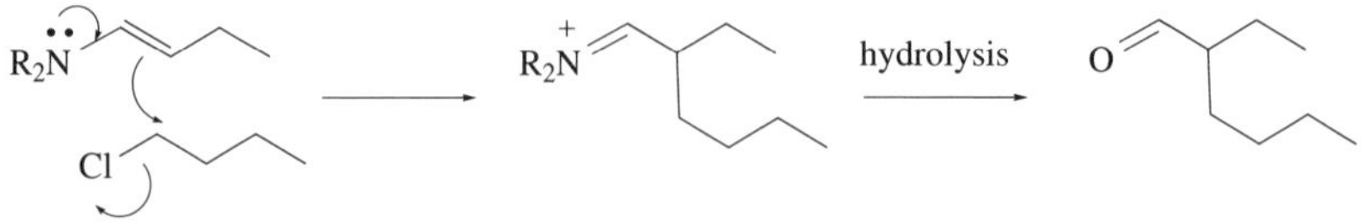

R_2N Cl $R_2\overset{+}{N}$ hydrolysis O

10.52 Addition of an Enolate to a Nitroso Group

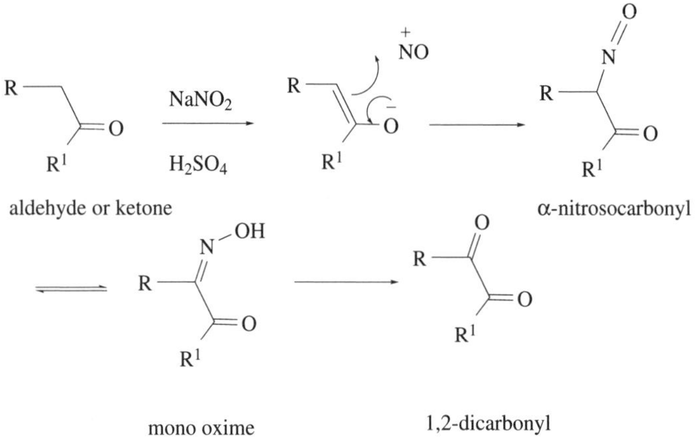

R R^1 O $NaNO_2$ H_2SO_4 R R^1 $\overset{-}{O}$ $\overset{+}{NO}$ R N O R^1 O

aldehyde or ketone α-nitrosocarbonyl

R N OH O R^1 R O O R^1

mono oxime 1,2-dicarbonyl

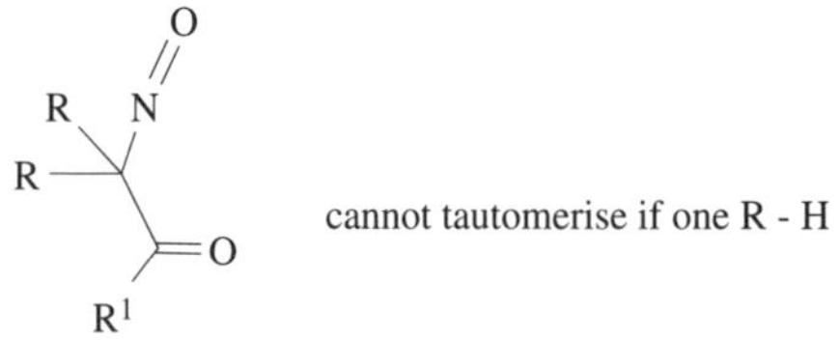

R R N O O R^1

cannot tautomerise if one R - H

This is a method of producing 1,2-dicarbonyl compounds. Selenium dioxide in ethanoic (acetic) acid also produces 1,2-dicarbonyl compounds.

$$R^1CH_2C(=O)R^2 \xrightarrow[CH_3CO_2H]{SeO_2} R^3C(=O)C(=O)R^4$$

$R^1, R^2 = H$	80% yield
$R^1, R^2 = CH_3$	17% yield, R^3 = -CH_2CH3 R^4 = H. Me oxidised quicker than CH_2

The yields with this reagent are often not very good.

10.53 Preparation of Carboxylic Acids by Oxidation of Alcohols or Aldehydes

$$RCH_2OH \xrightarrow{\text{oxidation}} RCHO \xrightarrow{\text{oxidation}} RCO_2H$$

If the oxidation starts with the alcohol it proceeds cleanly to the carboxylic acid. If the aldehyde is used as the starting material and is in significant concentration then there is formation of some aldol products.

10.54 Preparation of Carboxylic Acids by Reaction of a Grignard with Solid Carbon Dioxide

$$RMgBr \xrightarrow{\text{solid } CO_2} RC(=O)OMgBr \xrightarrow{H^+/H_2O} RC(=O)OH$$

10.55 Preparation of Carboxylic Acids by Hydrolysis of Nitriles

$$RCN \xrightarrow[\text{or } HO^-]{H^+/H_2O} RC(=O)NH_2 \xrightarrow{H^+/H_2O} RC(=O)OH$$

10.56 Preparation of Carboxylic Acids by the Haloform Reaction

See 10.44 and 10.45.

$$RCOCH_3 \xrightarrow[NaOH]{Cl_2} RCOO^- Na^+ + HCCl_3$$

10.57 Preparation of Carboxylic Acids by Conversion of Trihalides

See 11.30.

$$RCX_3 \xrightarrow{NaOH} RCOO^- Na^+ \xrightarrow{H^+/H_2O} RCOOH$$

10.58 Dissociation Constants of Carboxylic Acids

All carboxylic acids possess a dissociation constant.

$$RCOOH + H_2O \rightleftharpoons RCOO^- + H_3O^+$$

$$HA\ (+\ H_2O) \rightleftharpoons H^+ + A^-$$

$$\text{acid dissociation constant } K_a = \frac{[H^+][A^-]}{[HA]}$$

$$pK_a = -\log_{10} K_a$$

The pK_a of any compound is a measure of how difficult it is to remove the proton whose pK_a is being quoted. The —**OH** proton of a carboxylic acid has a pK_a of 4.75. The —**OH** proton of an aliphatic alcohol has a pK_a of 18. It is a lot easier for a given base to remove the proton from a carboxylic acid than from an aliphatic alcohol, or in other words carboxylic acids are stronger acids relative to alcohols.

There are two types of stabilisation occuring in the equilibrium equation above. There is stabilisation of the anion by resonance:

$$RC(=O)O^- \longleftrightarrow RC(O^-)=O$$

and there is stabilisation of the carboxylic acid by resonance:

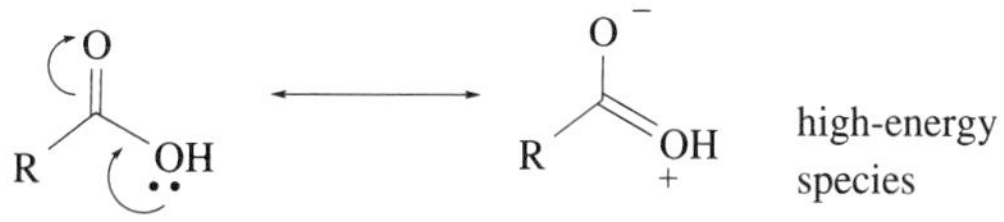

$$RC(=O)\ddot{O}H \longleftrightarrow RC(O^-)=\overset{+}{O}H \quad \text{high-energy species}$$

The first type of stabilisation is more important than the second. The position of the equilibrium depends to a large extent on the stabilisation of the anion. Anything which stabilises this anion pulls the equilibrium to the right and produces a smaller pK_a.

R	CH_3	CH_3CH_2	H	$H_3C—O—CH_2$	$N{\equiv}C—CH_2$	$ClCH_2$	Cl_2CH	Cl_3C
pK_a	4.75	4.87	3.75	3.5	2.4	2.75	1.3	0.6

10.59 Esterification of Carboxylic Acids Involving Mineral Acid Catalysis

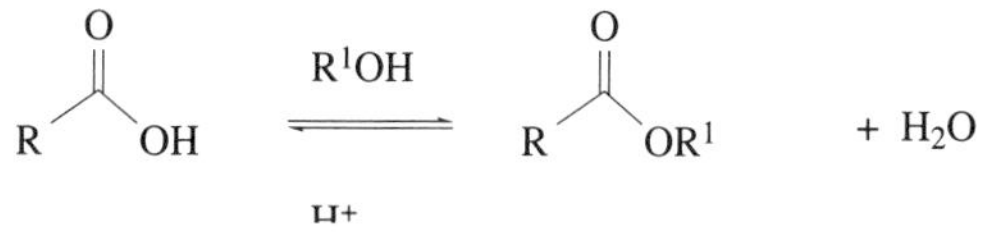

$$RC(=O)OH \underset{H^+}{\overset{R^1OH}{\rightleftharpoons}} RC(=O)OR^1 + H_2O$$

The water is removed as it is formed by adding benzene and distilling out the benzene/water azeotrope (note that for a hydrolysis of an ester a large excess of water is used by carrying out the reaction in an aqueous solution).

10.60 Mechanism of Esterification of Carboxylic Acids Involving Mineral Acid Catalysis

R–C(=O)–OH ⇌ R–C(=$\overset{+}{O}$H)–OH + R^1–ÖH ⇌ R–C(ÖH)(OH)–$\overset{+}{O}$(H)–R^1 (H^+, "base") ⇌ R–C($\overset{+}{O}H_2$)(OH)–O–R^1 ⇌ R–C(=O)–OR^1

10.61 Use of Acid Chlorides in Forming Esters

R–C(=O)–OH —($SOCl_2$)→ R–C(=O)–Cl —(R^1OH, pyridine)→ R–C(=O)–OR^1

pyridine

R–C(=O)–Cl + pyridine → R–C(=O)–$\overset{+}{N}C_5H_5$ + R^1–ÖH → R–C(=O)–OR^1

10.62 Esters from Silver Salts of Carboxylic Acids with an Alkyl Halide

R–C(=O)–OH —($AgNO_3$)→ R–C(=O)–$\overset{-}{O}$ $\overset{+}{Ag}$ —(reflux, R^1I)→ R–C(=O)–OR^1 + AgI

10.63 Methyl Esters Using Diazomethane

Methyl esters can be produced using diazomethane as shown below. Diazomethane is a powerful electrophile. It will also react with acid chlorides; this is known as the Arndt–Eistert reaction.

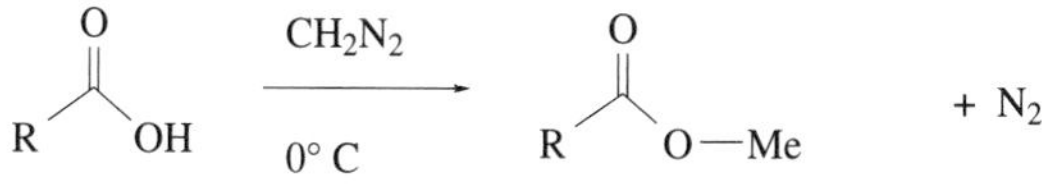

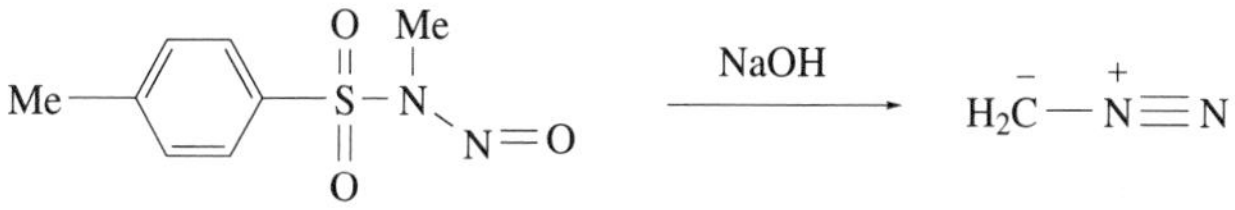

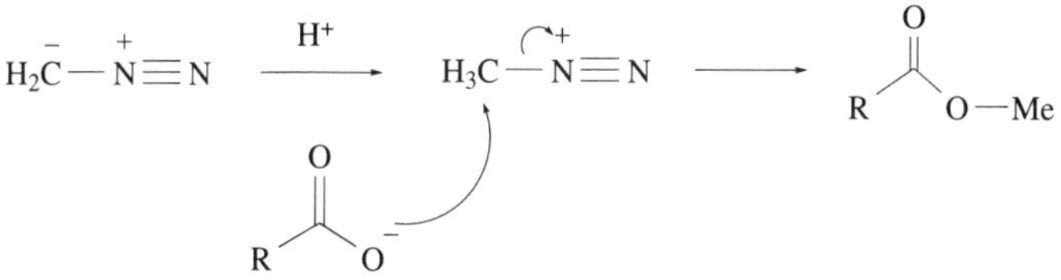

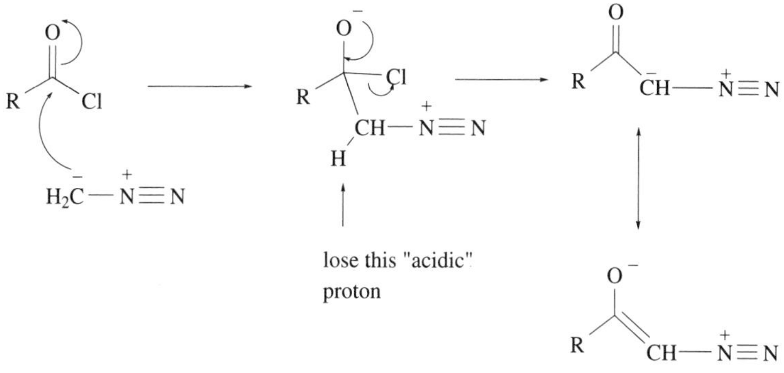

$$RCOOH \xrightarrow[0^\circ C]{CH_2N_2} RCO{-}O{-}Me + N_2$$

$$Me{-}C_6H_4{-}SO_2{-}N(Me){-}N{=}O \xrightarrow{NaOH} H_2\overset{-}{C}{-}\overset{+}{N}{\equiv}N$$

$$H_2\overset{-}{C}{-}\overset{+}{N}{\equiv}N \xrightarrow{H^+} H_3C{-}\overset{+}{N}{\equiv}N \longrightarrow RCO{-}O{-}Me$$

$RCOO^-$

$$RCOCl + H_2\overset{-}{C}{-}\overset{+}{N}{\equiv}N \longrightarrow R{-}C(O^-)(Cl){-}CH(H){-}\overset{+}{N}{\equiv}N \longrightarrow RCO{-}\overset{-}{C}H{-}\overset{+}{N}{\equiv}N \longleftrightarrow R{-}C(O^-){=}CH{-}\overset{+}{N}{\equiv}N$$

lose this "acidic" proton

10.64 The Wolff Reaction

$$RC(=O)\overset{-}{C}H-\overset{+}{N}\equiv N \xrightarrow{H_2O,\ Ag_2O} RCH_2C(=O)OH$$

acyl carbene

$$R-CH=C=O \quad (H^+,\ H_2\ddot{O})$$

The Wolff reaction (part of the Arndt–Eistert synthesis) is a method for inserting a methylene group between the carbonyl and the alkyl/aryl group in the carboxylic acid. The diazoketone is formed from the acid chloride when the acid chloride is not in excess.

$$RC(=O)OH \longrightarrow RCH_2C(=O)OH$$

$$RC(=O)R \xrightarrow[H^+]{CH_2N_2} RCH_2C(=O)R$$

10.65 Base Hydrolysis of Esters

$$RCOOR^1 \xrightarrow[H_2O]{NaOH} RCOO^- Na^+ + R^1OH$$

10.66 Kolbe Electrolysis of Acid Salts

$$RCOO^- K^+ \xrightarrow{\text{electrolytic oxidation}} RCOO\bullet \longrightarrow R\bullet + CO_2 \longrightarrow R{-}R$$

This is a method for building long hydrocarbon chains.

10.67 Hunsdieker Reaction to Form Alkyl Halides

$$RCOO^- Ag^+ \xrightarrow[Br_2]{CCl_4} RCOO{-}Br \longrightarrow RBr + CO_2$$

$$RCOO{-}Br \longrightarrow R\bullet + Br\bullet + CO_2 \longrightarrow RBr + CO_2$$

10.68 Decarboxylation of Carboxylic Acids

In general carboxylic acids are thermally stable up to about 700 °C. There are, however, carboxylic acids that lose carbon dioxide at lower temperatures.

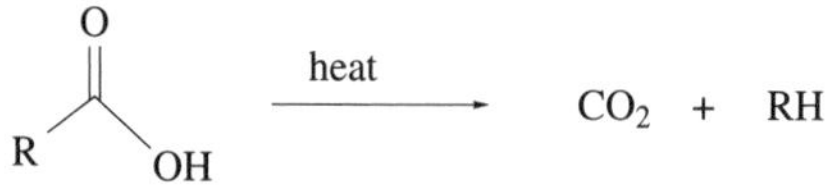

RCOOH —heat→ CO_2 + RH

Carboxylic acids of the type below readily lose carbon dioxide at around 100 °C.

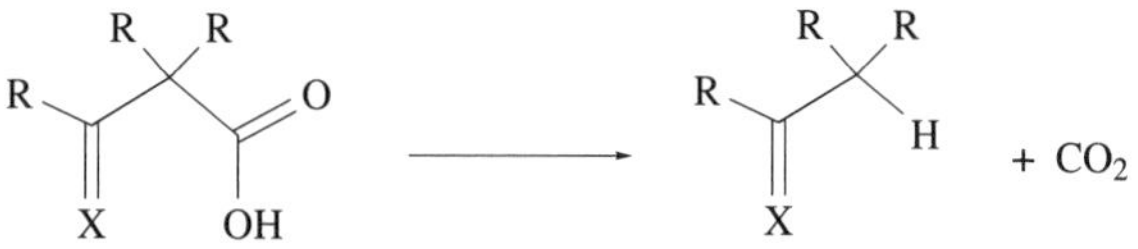

X = C (β,γ-unsaturated acids) or O (β-ketoacids)

10.69 Mechanism of Decarboxylation of Carboxylic Acids

The reaction goes *via* a six-membered transition state.

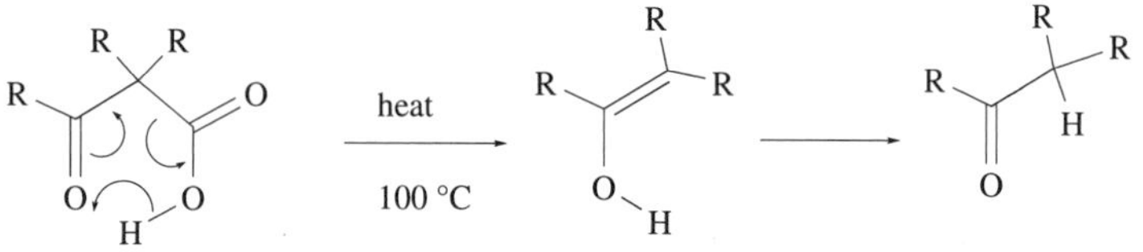

heat, 100 °C

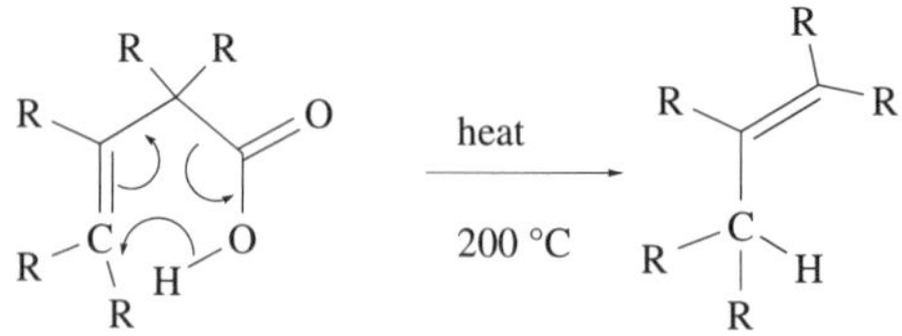

heat, 200 °C

A variation allows bromine to replace the carboxylic group.

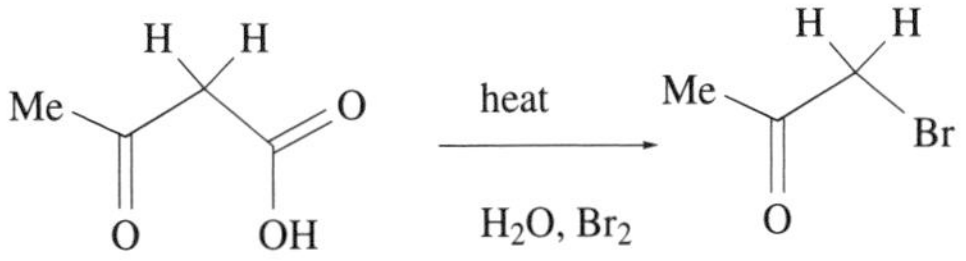

Me H H O OH O heat H_2O, Br_2 Me H H Br O

If X is carbon, then higher temperatures (180 °C) are required.

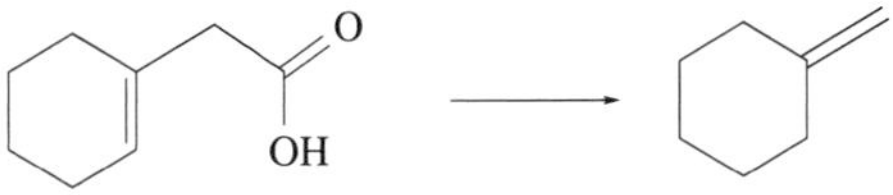

O OH

exocyclic C-C double bond

10.70 Decarboxylation of α,β-Unsaturated Carboxylic Acids

These can be decarboxylated at 200 °C. The reaction proceeds through the β,γ-unsaturated acid.

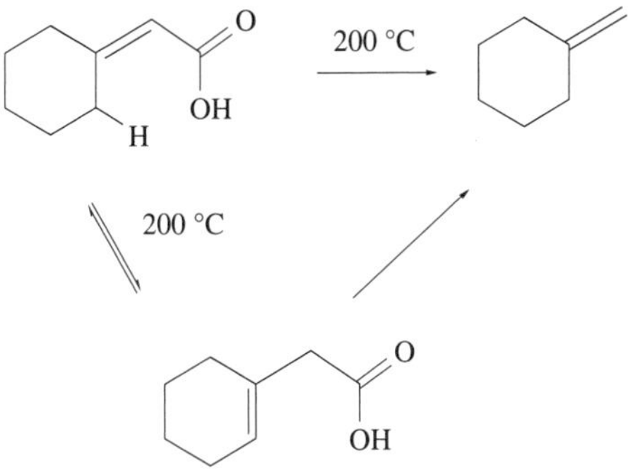

O OH H 200 °C

200 °C

O OH

10.71 Decarboxylation of Anions of Carboxylic Acids

O R OH

stable

O R O −

unstable

O R O^- CO_2 + R^-

10.72 Decarboxylation of Halogen-substituted Carboxylic Acids

It is possible to prepare nitromethane *via* decarboxylation of a halogen-substituted acid.

$$\mathrm{ClCH_2C\overset{-}{O_2}\ \overset{+}{Na}} \xrightarrow[\text{heat}]{\mathrm{NaNO_2}} \mathrm{O_2NCH_2C\overset{-}{O_2}\ \overset{+}{Na}} \xrightarrow{\text{heat}} \mathrm{\overset{-}{C}H_2NO_2}$$

$$\xrightarrow{\mathrm{H^+}} \mathrm{CH_3NO_2}$$

Other halogen-substituted acids such as trichloroethanoic acid give chloroform as the product.

$$\mathrm{Cl_3CC\overset{-}{O_2}\ \overset{+}{Na}} \xrightarrow[\text{reflux}]{\mathrm{H_2O}} \mathrm{CHCl_3}$$

The mechanism for this reaction is as follows:

good leaving group

X(HO)C=O ——(alkaline solution)——→ X(⁻O)C=O ——→ $\bar{X}$ + CO_2

10.73 Decarboxylation of Dicarboxylic Acids

$R_2C(CO_2H)–C(CO_2H)R_2$ ——$Pb(OAc)_4$——→ $R_2C(CO_2^{\bullet})–C(CO_2^{\bullet})R_2$ ——→ $R_2C=CR_2$

10.74 The Hell-Volhard-Zelinskii Reaction

The α-hydrogens of carboxylic acids can be replaced by bromine or chlorine with a phosphorus halide catalyst. When there are two α-hydrogens, both may be replaced. The reaction takes place *via* the acid halide. See also 10.45 and 10.46.

$$RCH_2CO_2H \xrightarrow{Br_2} RCHBrCO_2H \xrightarrow{Br_2} RCBr_2CO_2H$$

10.75 Reactions of Esters: Reduction

$$RCO{-}OR^1 \xrightarrow{LiAlH_4} RCH_2OH$$

10.76 Condensation Reactions of Esters

See 10.48.

$$RCO{-}OR^1 \xrightarrow[2.\ H^+/H_2O]{1.\ Na,\ ether} RCH(OH)COR^1$$

an acyloin

$$RC(O^-)^{\bullet}{-}OR^1 \longrightarrow RC(O^-)(OR^1){-}C(O^-)(OR^1)R \longrightarrow RCOCOR^1 \xrightarrow{2e^-} R(^-O)C{=}C(O^-)R^1 \xrightarrow{H^+} RCH(OH)COR^1$$

(first step: $RCO{-}OR^1 \xrightarrow{e^-} RC(O^-)^{\bullet}{-}OR^1$)

$$EtO_2C(CH_2)_4CO_2Et \longrightarrow \text{2-hydroxycyclohexanone}$$

10.77 Reaction of Esters with Hydrazine

$$RCO{-}OR^1 \xrightarrow{H_2N-X} RCONHX + R^1OH$$

X = H (NH_3), R (alkyl, aryl amines), NH_2 (hydrazine)

10.78 The Curtius Rearrangement of Acyl Azides

$$RC(=O)O{-}R^1 \xrightarrow{H_2N{-}NH_2} RC(=O)NHNH_2 \text{ (hydrazide)} \xrightarrow[H_2SO_4,\ 0^\circ C]{NaNO_2} RC(=O)N_3 \text{ (acyl azide)}$$

$$RC(=O)\overset{-}{N}{-}\overset{+}{N}{\equiv}N \longrightarrow R{-}N{=}C{=}O \text{ (isocyanate)} + N_2$$

$$\xrightarrow{H_2O} RNHC(=O)OH \xrightarrow{R{-}N{=}C{=}O} RNHC(=O)NHR$$

$$RNHC(=O)OH \longrightarrow RNH_2$$

10.79 Industrial Preparation of Amides

Industrial preparation of amides often involves heating the ammonium salts of carboxylic acids.

$$CH_3CH_2C(=O)\overset{-}{O}\ \overset{+}{N}H_4 \xrightarrow{\text{heat}} CH_3CH_2C(=O)NH_2$$

10.80 Laboratory Preparation of Amides

Amides are prepared in the laboratory by the reaction of ammonia or amine with acid chlorides or with acid anhydrides.

$$CH_3CH_2C(=O)Cl \xrightarrow{R_2NH} CH_3CH_2C(=O)NR_2$$

$$CH_3C(=O)OC(=O)CH_3 \xrightarrow{R_2NH} CH_3C(=O)NR_2$$

10.81 The Hofmann Reaction/Rearrangement: Degradation of Amides

This is the conversion of primary amides into amines containing one carbon less.

$$\mathrm{RC(=O)NH_2} \xrightarrow[\mathrm{KOH}]{\mathrm{Br_2}} \mathrm{RNH_2}$$

$$\mathrm{RC(=O)NH_2} \longrightarrow \mathrm{RC(=O)NHBr} \longrightarrow \mathrm{RC(=O)N^-Br} \longrightarrow \mathrm{R{-}N{=}C{=}O} \xrightarrow{\mathrm{H_2O}} \mathrm{RNHC(=O)OH} \longrightarrow \mathrm{RNH_2}$$

isocyanate

or *via* the acyl nitrene

$$\left[\mathrm{RC(=O)\ddot{N}\!:}\right] + \mathrm{Br^-}$$

10.82 Hydrolysis of Amides

Hydrolysis of amides involves nucleophilic substitution in which ammonia or amines are produced when the nitrogen-containing fragment of the amide is displaced by a hydroxyl group. Under acidic conditions hydrolysis involves attack by water on the protonated amide and under alkaline conditions hydrolysis involves attack by the strongly nucleophilic hydroxide ion on the amide.

$$\mathrm{CH_3CH_2C(=O)NR_2} \xrightarrow{\mathrm{H^+/H_2O}} \mathrm{CH_3CH_2C(=O)OH} + \mathrm{NH_4^+}$$

$$\mathrm{CH_3CH_2C(=O)NR_2} \xrightarrow{\mathrm{HO^-/H_2O}} \mathrm{CH_3CH_2C(=O)O^-} + \mathrm{NH_3}$$

10.83 Conversion of Amides to Imides

Imides are the nitrogen analogue of anhydrides. They can be formed intermolecularly from dehydration of carboxylic acid–amide mixes. They can also be formed intramolecularly.

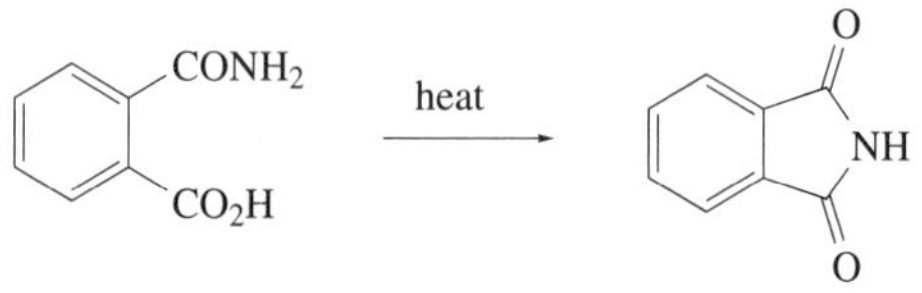

$CONH_2$ / CO_2H —heat→ phthalimide

10.84 Preparation of Acid Chlorides from $SOCl_2$ and Carboxylic Acids

Acid chlorides may be produced from the reaction of thionyl chloride with a carboxylic acid. This is a convenient method as the side products formed are gases, and any excess thionyl chloride (low boiling) may be removed by distillation.

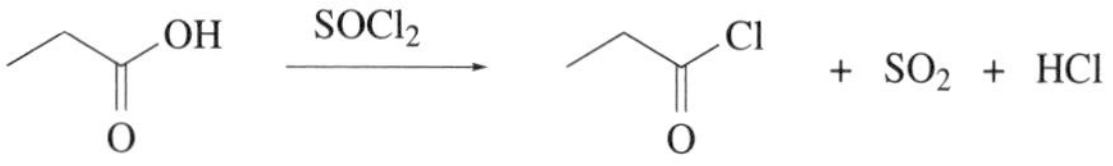

$SOCl_2$ → $+ SO_2 + HCl$

10.85 Preparation of Acid Chlorides from PCl_3 and Carboxylic Acids

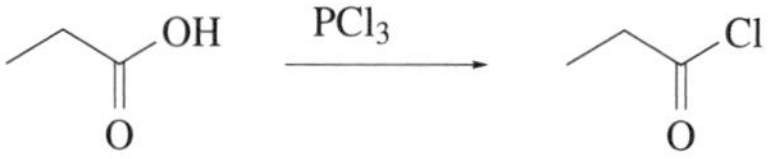

PCl_3

10.86 Preparation of Acid Chlorides from PCl_5 and carboxylic Acids

PCl_5

10.87 Conversion of Acid Chlorides to Acids

Acid chlorides can be converted back to the parent carboxylic acid by hydrolysis.

$$CH_3CH_2COCl \xrightarrow{H_2O} CH_3CH_2COOH$$

10.88 Conversion of Acid Chlorides to Amides

See 10-80.

10.89 The Schotten–Baumann Reaction: Conversion of Acid Chlorides to Esters

Esters may be formed from acid chlorides. Alkyl acid chlorides are reactive and often react easily with alcohols to give the ester. Aromatic acid chlorides are a lot less reactive and different conditions are required to form esters: the Schotten–Baumann reaction. The acid chloride is added in portions to a mixture of the hydroxy compound and a base, usually aqueous sodium hydroxide or pyridine, see 10.61. The base serves to catalyse the reaction and to remove the HCl produced.

10.90 The Friedel–Crafts Reaction: Conversion of Acid Chlorides to Ketones

See 11.17 and 10.20.

10.91 Organocopper Compounds in the Conversion of Acid Chlorides to Ketones

Organocopper reagents, formed from alkyl lithiums and copper salts, react with aryl or alkyl acid halides to give the corresponding ketone.

$$CH_3CH_2COCl \xrightarrow{R_2CuLi} CH_3CH_2COR$$

10.92 Formation of Aldehydes from Acid Chlorides by Reduction

Aryl or alkyl acid halides can be reduced using lithium tri-*tert*-butoxyalumino hydride to give aldehydes.

$$CH_3CH_2COCl \xrightarrow{LiAlH(O^tBu)_3} CH_3CH_2CHO$$

10.93 Preparation of Anhydrides Using Ethanoic Acid with a Ketene

The most commonly encountered monocarboxylic acid anhydride is ethanoic (acetic) anhydride. It is prepared by the reaction of ethanoic acid with the ketene $CH_2{=}C{=}O$, itself prepared by the high-temperature dehydration of ethanoic acid.

$$CH_3COOH \xrightarrow{AlPO_4} H_2O + H_2C{=}C{=}O \text{ (ketene)} \xrightarrow{CH_3COOH} (CH_3CO)_2O$$

10.94 Preparation of Anhydrides by Dehydration of Dicarboxylic Acids

Dicarboxylic acid anhydrides can be synthesised by simple heating if a five- or six-membered ring is the product.

$$C_6H_4(CO_2H)_2 \xrightarrow{heat} C_6H_4(CO)_2O$$

10.95 Conversion of Anhydrides to Carboxylic Acids

Simple hydrolysis of anhydrides with water gives the parent acids.

$$\text{phthalic anhydride} \xrightarrow{H_2O} C_6H_4(CO_2H)_2$$

10.96 Conversion of Anhydrides to Amines

Anhydrides react with ammonia or amines to give the corresponding amide. See 10.80.

10.97 Conversion of Anhydrides to Esters

Anhydrides react with alcohols to give esters.

$$\text{phthalic anhydride} \xrightarrow{EtOH} C_6H_4(CO_2Et)_2$$

10.98 The Friedel–Crafts Reaction: Formation of Ketones from Anhydrides

Like acid chlorides, anhydrides undergo the Friedel–Crafts reaction to form ketones.

$$(MeCO)_2O \; + \; ArH \longrightarrow MeC(O)Ar \; + \; RCO_2H$$

10.99 Relevant Questions

5, 19, 21, 34, 42(c), 45, 46, 54(ii), 57, 59, 62, 65, 76, 77, 84, 87, 100(a), 105, 118(a), 139, 144(a and b), 154, 155, 161(a), plus parts of functional group interconversion questions.

11 AROMATICITY: THE CHEMISTRY OF BENZENE AND DERIVATIVES

11.1 Aromaticity

The core aromatic compounds tend to be very stable. For example, the combustion of benzene gives 3300 kJ mol^{-1} of energy. That of the hypothetical 1,3,5-cyclohexatriene would be 3448 kJ mol^{-1}. Benzene is 148 kJ mol^{-1} more stable. The difference is **delocalisation** or **resonance** energy.

The chemical reactivity of aromatic compounds shows (i) reduced unsaturation, (ii) substitution rather than addition reactions, and (iii) stability to heat and light.

The nuclear magnetic spectra and properties of aromatic compounds are also characteristic. Aromatic compounds have ring currents. The systems are **diatropic** (having the ability to sustain a ring current not caused by unpaired electrons). The magnetic field resulting from a diatropic system is **diamagnetic**. **Paratropic** systems are those which can sustain a ring current due to unpaired electrons. The magnetic field resulting from this is known as **paramagnetic**.

This leads to a definition of aromaticity:

> "A closed shell of delocalised electrons and resultant ability to sustain an induced ring current."

11.2 Hückel's Rule

Amongst fully conjugated, planar, cyclic, delocalised molecules, those containing $4n + 2$ (n = an integer) electrons will have special "aromatic character".

11.3 Aromaticity: Cyclopentadiene

The orbital diagrams for the cation, radical and anion of cyclopentadiene show that the anion is the most stable as it has its orbitals full.

The stability cannot arise from classical delocalisation energy as all three of these systems have canonical forms. In fact, the anion conforms to Hückel's $(4n + 2)$ rule and has aromatic character.

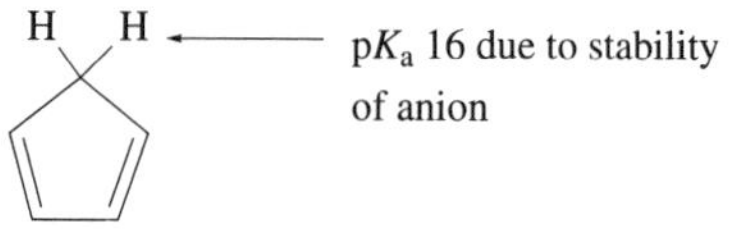

H H

pK_a 16 due to stability of anion

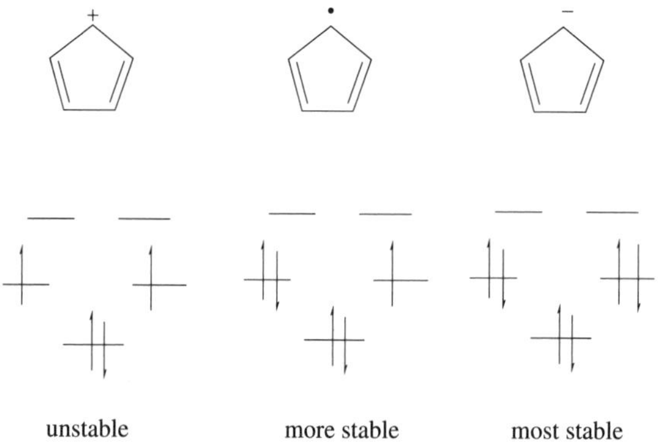

+ • −

unstable more stable most stable

The structure of the cyclopentadiene anion is a planar, regular pentagon. It has an estimated 173 kJ mol^{-1} of delocalisation energy. It is, like benzene, an unusually stable system. However, its chemical reactivity is not like that of benzene: it reacts with electrophiles, MeI, CO_2, *etc.* Cyclopentadiene also shows a diamagnetic ring current.

11.4 Aromaticity: Six-electron Seven-membered Rings and Tropone

The tropylium ion or cycloheptatrienyl cation also shows aromatic character. Tropone exists almost exclusively as the ylide, as this is a highly stable aromatic system.

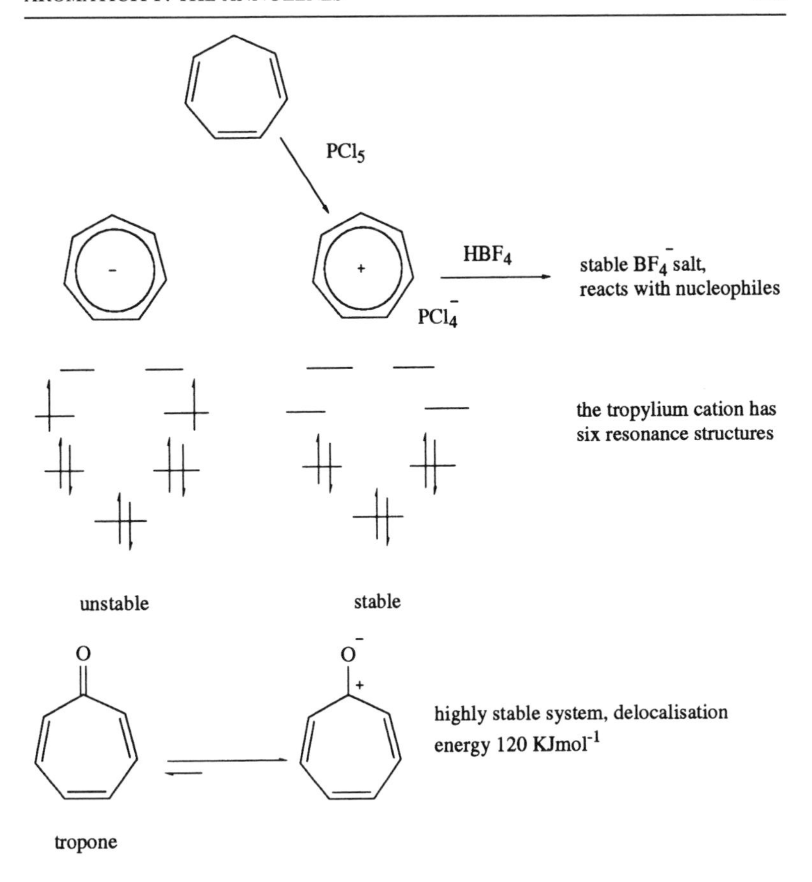

11.5 Aromaticity: the Annulenes

The properties of the annulenes are summarised in the following table.

Number of pi electrons	Annulene	Comments (δ = NMR shift)
2	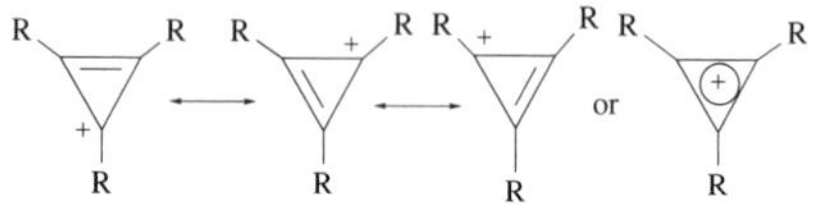 or	ring current, diatropic, when R = H, δ= 11.1
4		anti-aromatic, an unusually stable 4*n* system
6		see 11.7
8	  "tub"	non-aromatic, behaves as a poly-olefin, tub-shaped as there is no advantage in being planar
10	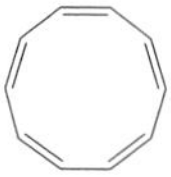 (i) all *cis*  (ii) mono *trans* 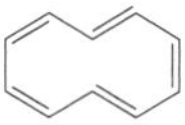 (iii) *cis-trans-cis-cis-trans*	(i) unstable, enormous bond angle strain (ii) very reactive (iii) non-planar, two inner hydrogens interfere
10	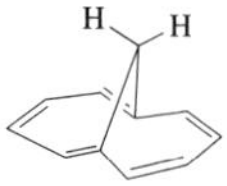 bridged	diatropic, bridgehead hydrogens have δ = -0.5, planar, bond equivalency
12		paratropic 4n system, paramagnetic ring current at -170 °C, δ = 8.0 for inner hydrogens and 6.0 for outer
14	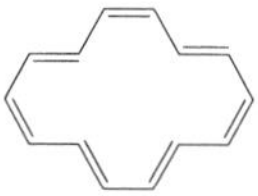	diatropic 4*n*+2 system, at -130 °C outer hydrogens have δ = 7.6 and inner δ = 0
16	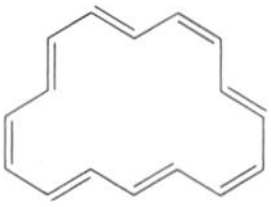	paratropic 4*n* system, at -130 °C outer hydrogens have δ = 5.3 and inner δ = 10.5
18	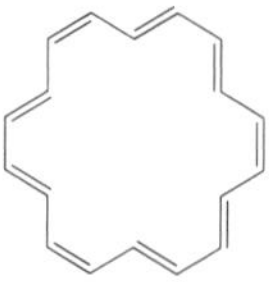	diatropic, diamagnetic system, at -60 °C outer hydrogens have δ = 9.3 and inner δ = -3

11.6 The Concept of Anti-aromaticity

If a planar, cyclic conjugated molecule is destabilised relative to its acyclic, but conjugated, counterpart, then it is considered to have **anti-aromaticity**. Most examples of this tend to occur in small rings.

Note that only the essentials of aromaticity have been covered here, many courses will include *Condensed Aromatic Systems* under this heading. If these are covered in your course you should know something about the following topics: general properties and reactions of naphthalene, anthracene and phenanthrene (including bond alternation (fixation), activating/deactivating substituents on naphthalene, bromination (and other electrophiles), hydrogenation and oxidation of phenanthrene, Diels–Alder of phenanthrene to produce tryptycene); synthesis of naphthalene, anthracene and phenanthrene. Other systems including tetracene, coronene and helicenes.

11.7 Benzene: General Background

Benzene is represented by the Kekulé structure:

There are other representations of benzene or the phenyl group; these are shown below.

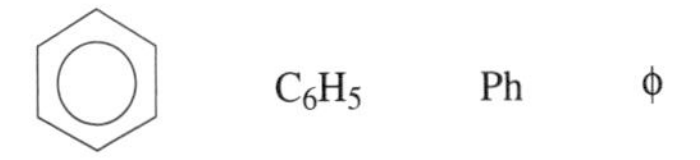

Benzene has a high stability to heat and oxidation and undergoes substitution rather than addition reactions. This is illustrated by the following examples:

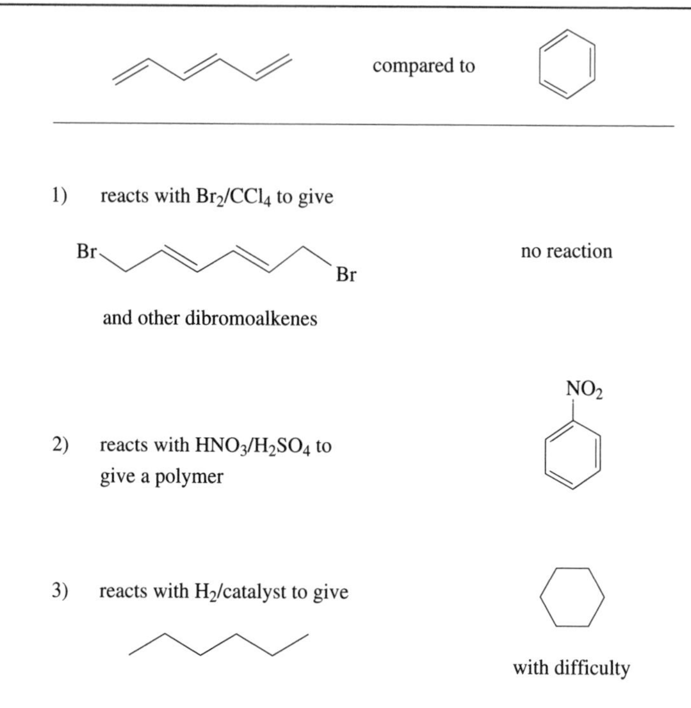

Functional groups attached to the ring can behave differently from those attached to aliphatic systems.

11.8 Structure of Benzene: Valence Bond Approach

The valence bond approach looks at the number of possible resonance canonicals and their contribution to the overall structure of the molecule.

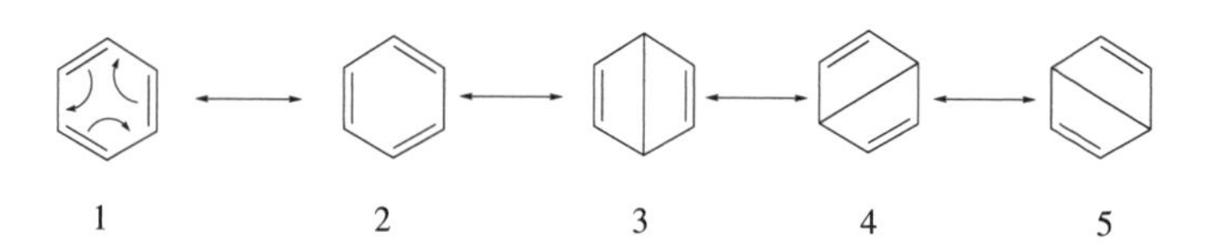

For benzene there is a 39% contribution each from 1 and 2, and 7.3% each from 3, 4 and 5. The resonance hybrid is more stable than any single resonance canonical.

11.9 Structure of Benzene: Molecular Orbital Approach

Here the structure of benzene is looked at in terms of its molecular orbitals. For benzene we take the six p orbitals available from the six sp^2 hybridised carbons and combine them to form six new molecular orbitals.

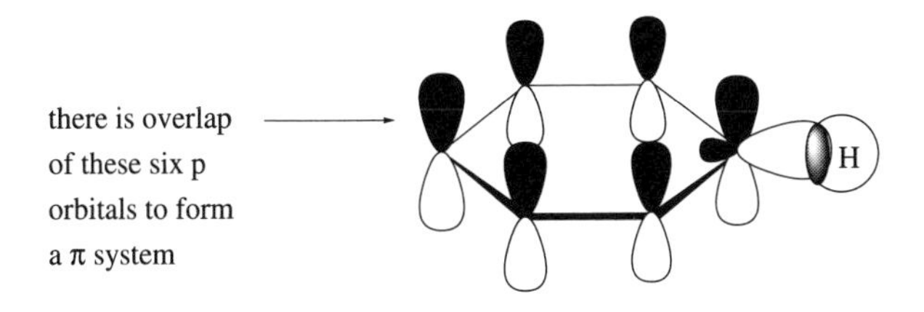
there is overlap
of these six p
orbitals to form
a π system
H
the carbons are sp² hybridised

six atomic p orbitals give six molecular orbitals of which three are bonding and three are antibonding

11.10 Reactions of Unsubstituted Benzene

There is benzene and there are substituted benzenes. There are two types of reaction associated with benzene and its substituted derivatives. These are electrophilic aromatic substitution reactions and nucleophilic substitution reactions.

11.11 Electrophilic Aromatic Substitution (EAS) of Benzene

In order to give benzene chemistry some structure, and hence allow predictions to be made in how substituted benzenes will behave in certain reactions, the way in which the substituents on benzene affect its behaviour is always related back to benzene; ***i.e.*** **benzene is the standard**.

The general mechanism for EAS of benzene is shown below. There then follow seven examples (11.12–11.18) of reactions that proceed by EAS.

known as a; Wheland intermediate
or arenium ion or benzenium ion
or σ-complex

The energy/reaction diagram below describes the changes in energy for the above process when $Y = Br^+$. It also compares the above process to that of bromination of a simple alkene.

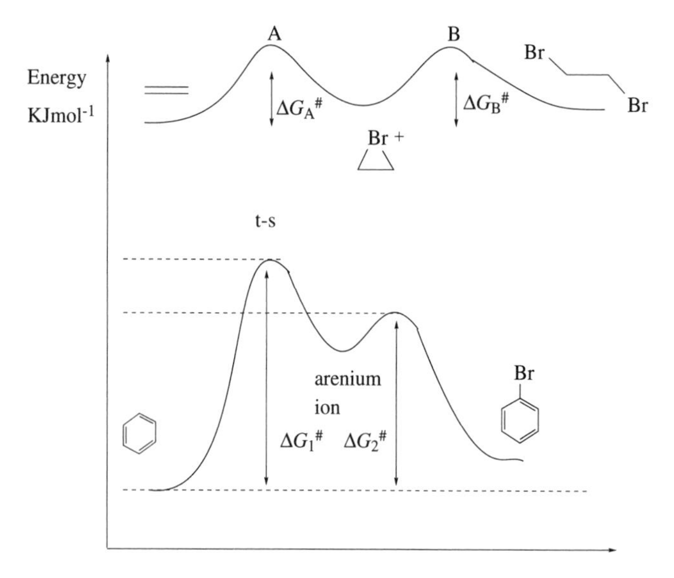

In the case of bromination of benzene $\Delta G_1^{\#}$ and $\Delta G_2^{\#}$ are much greater than $\Delta G_A^{\#}$ and $\Delta G_B^{\#}$ for either the formation of the bromonium ion (A) from the alkene or its subsequent attack (B) by Br^- to form the dibromoalkane. This is why a catalyst is needed for the bromination of benzene and not needed for the bromination of an alkene.

The intermediate is a delocalised cation and in most EAS reactions $K_1 > K_2$. Loss of H^+ is rapid and not an RLS. This sort of reaction can be thought of as addition followed by elimination. It is the relative energy of the transition state which determines $\Delta G^{\#}$. For an endothermic reaction there is a late transition state and therefore the transition state and the intermediate can be considered as one.

11.12 Protonation (Y = D or H) of Benzene

This is often used to form deuterated benzene. For deuteration D_2SO_4 is used. If excess D_2SO_4 is used hexa-substituted deuterated benzene is formed.

11.13 Halogenation (Y = Br, Cl) of Benzene

This reaction needs a halogen carrier, which is usually a Lewis acid. For bromination $FeBr_3/H^+$ or $HOBr/H^+$ are used. There are analogous reactions for chlorination. For iodination and fluorination indirect methods are normally used, see 11.18.

$$\text{C}_6\text{H}_6 + \text{Br}{-}\overset{+}{\text{O}}\text{H}_2 \rightleftharpoons [\text{C}_6\text{H}_6\text{Br}]^+ \longrightarrow \text{C}_6\text{H}_5\text{Br}$$

$$\text{C}_6\text{H}_6 + \overset{+}{\text{Cl}}\ \overset{-}{\text{AlCl}_4} \rightleftharpoons [\text{C}_6\text{H}_6\text{Cl}]^+\ \overset{-}{\text{AlCl}_4} \longrightarrow \text{C}_6\text{H}_5\text{Cl}$$

11.14 Nitration (Y = NO_2) of Benzene

HNO_3/H_2SO_4 are used to produce NO^{2+}. For substituted benzenes which are activated HNO_3 or HNO_3/HOAc will suffice.

$$\text{C}_6\text{H}_6 + \overset{+}{\text{NO}_2} \rightleftharpoons [\text{C}_6\text{H}_6\text{NO}_2]^+ \longrightarrow \text{C}_6\text{H}_5\text{NO}_2$$

$$HNO_3 + 2H^+ \rightleftharpoons NO_2^+ + H_3O^+$$

Unreactive compounds, such as benzoic acid and nitrobenzene, need HNO_3/H_2SO_4.

11.15 Sulfonation of Benzene

SO_3H is a removable “blocking group”.

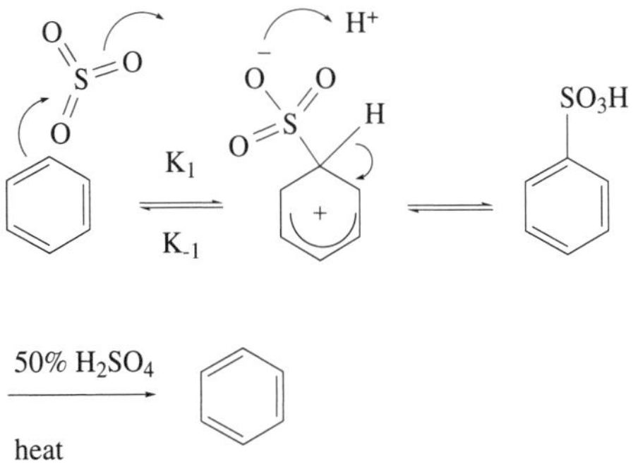

K_1

K_{-1}

SO_3H

H^+

50% H_2SO_4

heat

For this reaction K_{-1} approximately equals K_1, and therefore this is a reversible reaction.

11.16 Alkylation (Y = Alkyl) of Benzene

This reaction is known as the Friedel–Crafts alkylation reaction. It does not work for deactivated systems or for systems containing —OH, —OMe, NH_2, *etc.* as these coordinate to the Lewis acid catalyst. The alkyl halide is made electrophilic by coordination to a Lewis acid such as an aluminium halide, AlX_3. The intermediate complex is, for example, $CH_3CH_2{-}Br^+{-}AlBr_3{-}$.

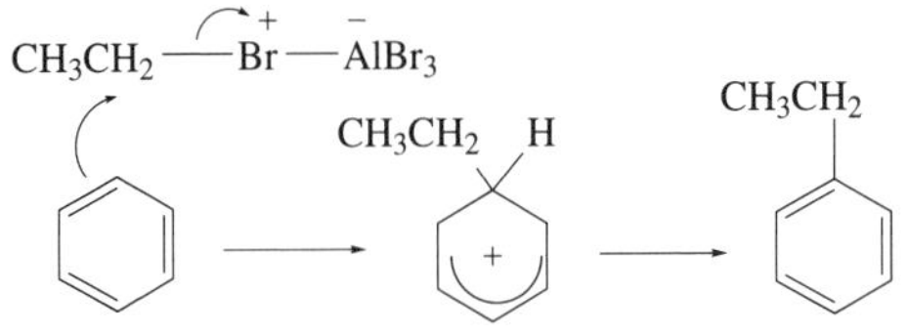

CH_3CH_2—$\overset{+}{Br}$—$\overset{-}{AlBr_3}$

CH_3CH_2 H

CH_3CH_2

Problems arise with multialkylation and rearrangement.

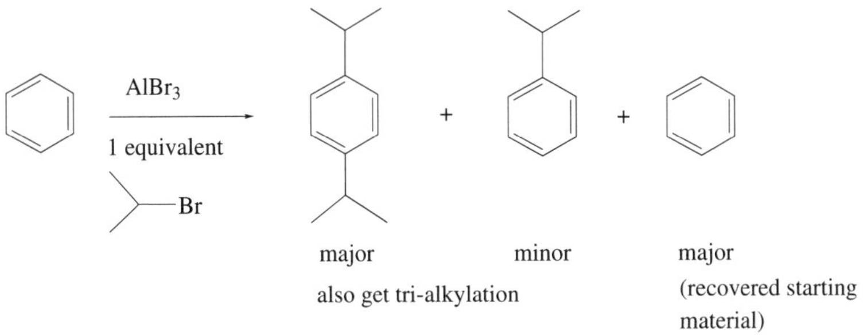

$AlBr_3$

1 equivalent

Br

major

minor

major

also get tri-alkylation

(recovered starting material)

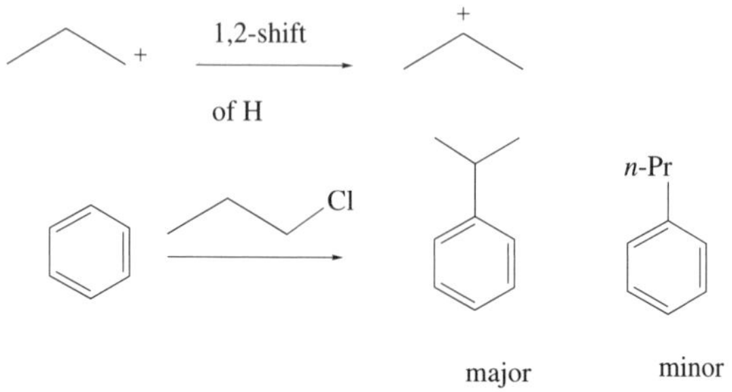

1,2-shift

of H

Cl

n-Pr

major minor

There are other methods using alkenes or alcohols with acid.

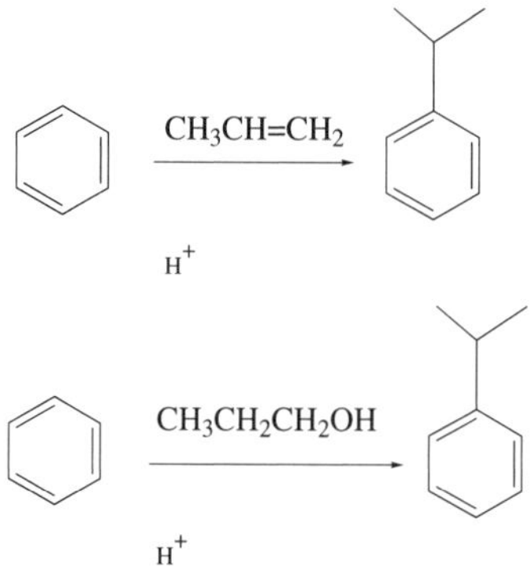

$CH_3CH{=}CH_2$

H^+

$CH_3CH_2CH_2OH$

H^+

11.17 Acylation (Y = RCO) of Benzene

This reaction is known as the Friedel–Crafts acylation reaction. It is basically the same as above but acyl halides are utilised instead of alkyl halides. The intemediate complex is, for example, $CH_3CO–Cl^+–AlCl_3–$; here Y = $CH_3CO–$.

R–C(=O)Cl —$AlCl_3$→ R–C(=O)–Cl$^+$–$AlCl_3^-$ ⇌ [R–C$^+$=Ö ↔ R–C≡O^+] $AlCl_4^-$

acylium ion

No rearrangement is observed and we only get the monosubstitution product (the acyl group deactivates the benzene ring). It is an indirect method of attaching alkyl groups to benzene.

Benzene —RCOCl, $AlCl_3$→ PhCOR —reduction (Wolff–Kishner or Clemenson)→ $PhCH_2R$

A similar reaction takes place if carboxylic acid anhydrides are used in place of acyl chlorides.

11.18 Metallation ($Y = M^+$) of Benzene

Mercuric salts of benzene can be formed using $Hg(OAc)_2$.

Benzene —$Hg(OAc)_2$ (Hg^{2+})→ [$^+$Hg, H arenium ion] —slow→ Ph–Hg^+

Ph–Hg^+ —D_2SO_4→ PhD; —HX→ PhX; —other E^+, as long as they are very reactive→ PhE

Thallium salts can also be created using $Tl(O_2CCF)_3$ and are used to put fluorine into benzene; here Y = $Tl(O_2CCF)_2$. Reaction of the thallium complex with KF/BF_3 gives the fluorobenzene.

$Tl(O_2CCF_3)_3$ → $Tl(O_2CCF_3)_2$ (on benzene ring) → KF/BF_3 → F (on benzene ring)

11.19 Reactions of Substituted Benzenes

In order to make this subject simple it is possible to divide most of the possible groups that can be attached to the benzene ring into five categories. These are:

1. $+$**I**

This means the group has a **positive induction** effect, *i.e.* it pushes **electron density into** the benzene ring.

Examples: alkyl groups.

2. $-$**I**

This means the group has a **negative induction** effect, *i.e.* it pulls **electron density out** of the benzene ring.

Examples: —CCl_3, —CF_3, —$^+NR_3$.

3. $-$**I** and $-$**R**

This means the group has the properties of 2 above, and it also has a **resonance effect** which removes **electron density from** the benzene ring.

Examples. —NO_2, —SO_3H, —CO_2H, —$CONH_2$, —CHO, —COR, —SO_3R, —CO_2R, —CN.

Note: guideline for this type is any group with a carbon–heteroatom multiple bond.

4. $-$**I** (small effect) and $+$**R (large effect)**

This means the group has the properties of 2 above, and it also has a **resonance effect** which puts **electron density into** the ring.

Examples: —OH, —OR, —NH_2, —NR_2.

5. $-$**I** and $+$**R (effectively balanced)**. See halogen case below.

When a group has the overall ability to remove **electron density from** the ring it is said to be **deactivating**. This means that relative to benzene EAS reactions of the substituted benzene will be **slower**.

When a group has the overall ability to put **electron density into** the ring it is said to be **activating**. This means that relative to benzene EAS reactions of the substituted benzene will be **faster**.

11.20 Control of Regiochemistry: the Ability to Direct *o*-, *m*- and p- in Substituted Benzenes (X = Substituent, Y^+ = General Electrophile)

The diagram below illustrates the possibilities for resonance canonicals when the electrophile Y^+ ends up in the *ortho*, *meta* or *para* position relative to the substituent X. (Note that there is an alternative nomenclature to *ortho*, *meta* and *para*. This is the IUPAC numbering system, where the carbon with the substituent X on it is labelled number 1: *ortho* is then number 2 or 6, *meta* number 3 or 5, and *para* number 4. In other words the six carbons of the benzene ring are labelled 1 to 6, depending on the priorities of the substituents.)

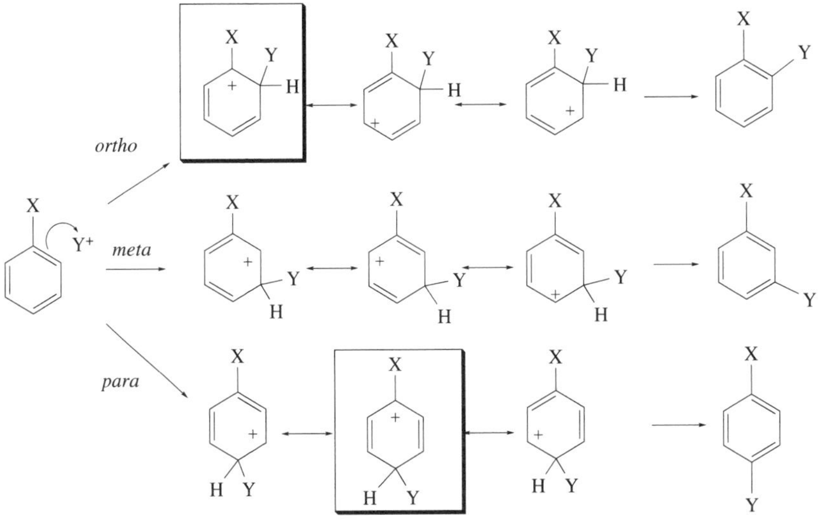

What we must consider is: how does the substituent X affect the course of the reaction? For the various types of substituent, the effects are:

1. **X** = +**I**

All arenium ions are stabilised relative to benzene; however, in the case of *o*- and *p*- the positive charge can sit directly on the carbon adjacent to the +**I** substituent and hence these two positions are favoured for substitution because the positive charge is stabilised by the input of electron density.

2. **X** = −**I**

All arenium ions are destabilised relative to benzene; however, in the case of *o*- and *p*- the positive charge can sit directly on the carbon adjacent to the −**-I** substituent and hence these two positions are less favoured because an already electron-deficient centre is losing more electron density for substitution, and hence *m*- becomes the favoured position.

3. **X** = −**I** and −**R**

Exactly the same as 2 above, but a much stronger effect.

4. **X** = −**I** and +**R**

Because the positive charge can be delocalised further onto the substituent when it sits on the carbon adjacent to the substituent the *o*- and *p*- positions are favoured. The arenium ions for these positions are highly stabilised. For the *m*- position the +**R** effect cannot operate and there is destabilisation at this position because of the −**I** effect of the substituent.

ortho

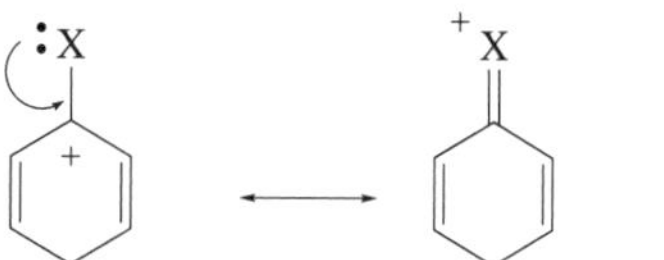

para

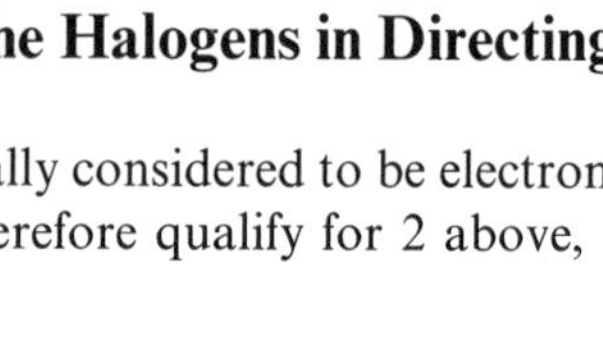

11.21 The Case of the Halogens in Directing *o*-, *m*- or *p*-

The halogens are generally considered to be electron-withdrawing substituents and should therefore qualify for 2 above, −**I**, and direct *m*-.

This is not the case. Halogens direct substitution reactions *o*- and *p*-. Here there is **deactivation** of the ring but *o*- and *p*- direction. This is because halogens carry a small +**R** effect. Although the *o*- and *p*-positions are destabilised by the −**I** effect of the halogen and the *m*-position is less destabilised, this changes when the slight +**R** effect comes into play. The situation is inverted and the *o*- and *p*- positions become slightly more favourable because the −**I** effect of the halogens is overridden. This +**R** stabilisation cannot occur if *meta* substitution takes place.

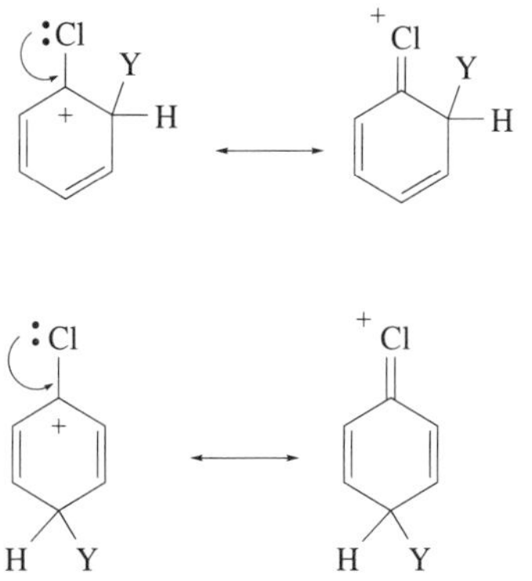

11.22 Directing Effects of Multisubstituted Benzenes

For multisubstituted benzenes, as an approximation, just add the effects of the individual groups to predict the directing effect of their combined effort. A guideline is that **activators** are more influential than **deactivators** and **resonance** is more important than **induction**. Do not forget steric effects.

Reinforcement:

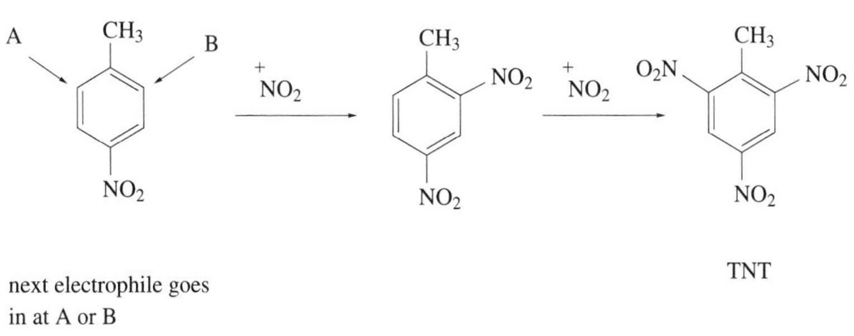

next electrophile goes in at A or B

TNT

Opposition:

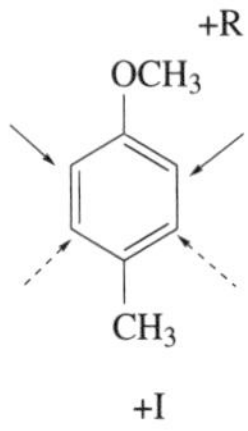

+R

OCH_3

CH_3

+I

next electrophile goes in at the carbons indicated by full arrows, +R effect dominates +I effect

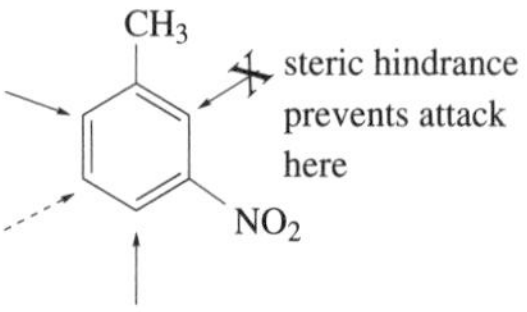

CH_3

steric hindrance prevents attack here

NO_2

the methyl group gives more stable intermediates, and the NO_2 group is deactivating

the next electrophile goes in as indicated by the full arrows

11.23 Important Reactions of Benzenes

In the reactions of benzene we aim for a single compound if possible, although separation of regioisomers (*o*-, *m*-, *p*-) is often possible.

11.24 Hydrogenation of Substituted Benzenes

Benzene with H_2/Pt, Rh or Ru:

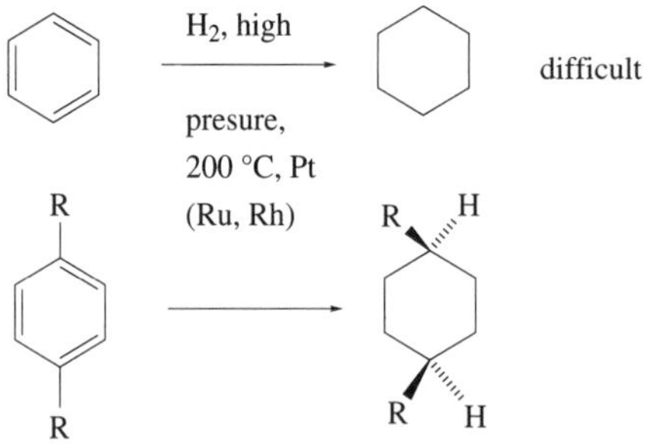

H_2, high presure, 200 °C, Pt (Ru, Rh)

difficult

R

R

R H

R H

11.25 Oxidation of Benzene and Substituted Benzene

Benzene with V_2O_5, $[O]/K_2Cr_2O_7/H^+$; 1,4-dimethyltoluene and V_2O_5/O_2:

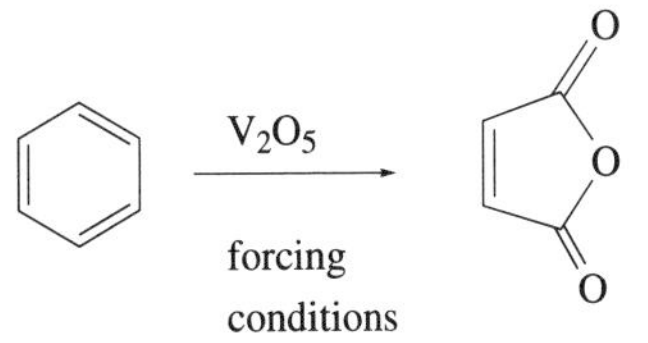

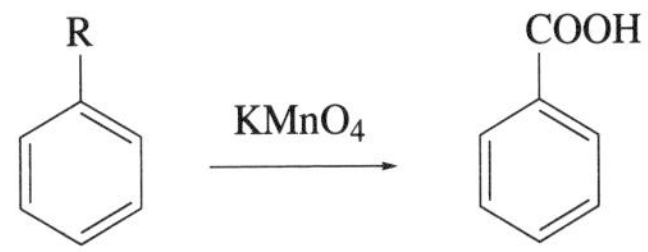

V_2O_5

forcing conditions

R

COOH

$KMnO_4$

can also use $K_2Cr_2O_7/H^+$

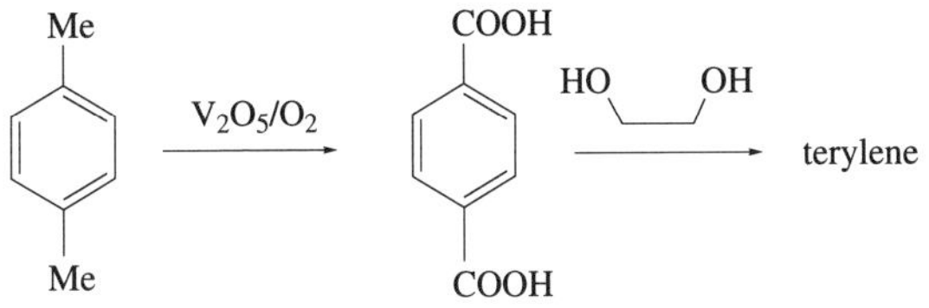

Me

Me

V_2O_5/O_2

COOH

COOH

HO OH

terylene

terephthalic acid

Me

Me

oxidation

heat

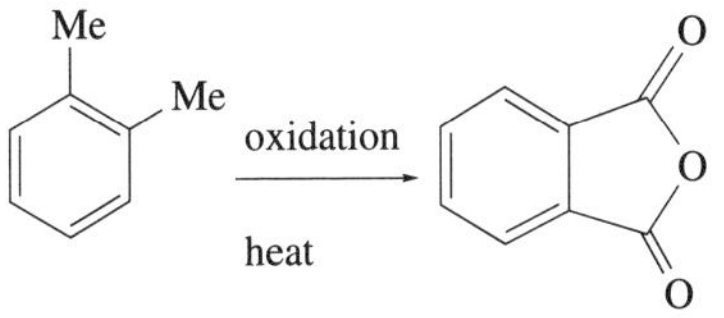

phthalic anhydride

11.26 Chlorination of Benzene

Benzene can be chlorinated with Cl_2 and $FeCl_3$ to give the monosubstituted benzene or with $Cl_2/h\nu$ to give the hexasubstituted benzene.

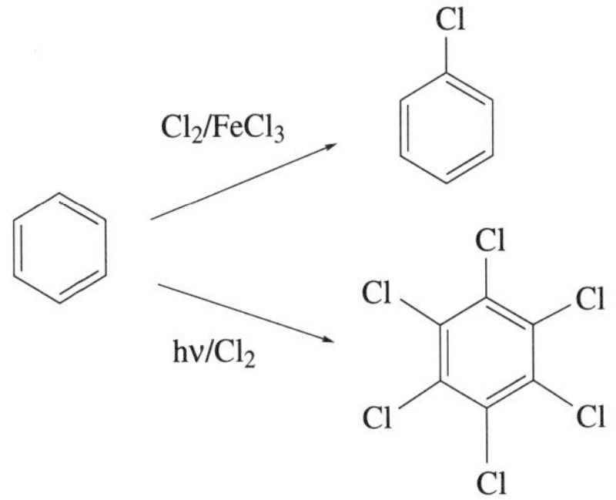

$Cl_2/FeCl_3$

Cl

$h\nu/Cl_2$

Cl Cl Cl Cl Cl Cl

11.27 Chlorination of Bromobenzene

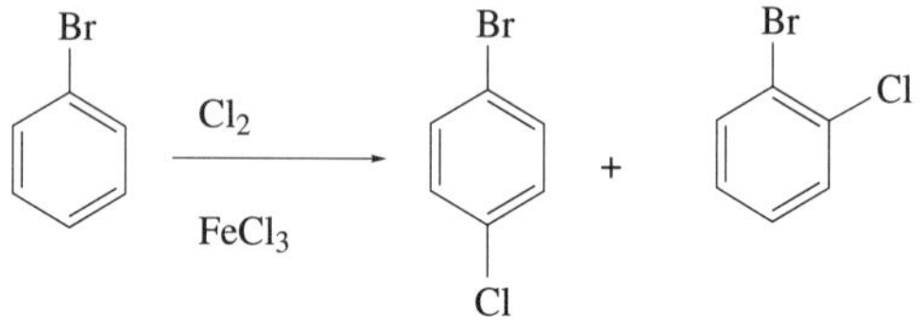

Br Br Br
Cl
Cl_2
+
$FeCl_3$
Cl

11.28 Benzene Side-chain Halogenation

The side-chain of a benzene derivative, such as methylbenzene (toluene) can be chlorinated using $Cl_2/h\nu$ or *N*-chlorosuccinimide (NCS).

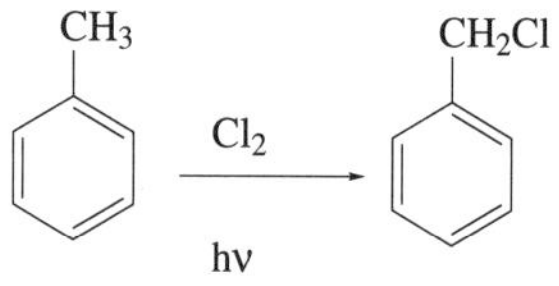

CH_3 CH_2Cl
Cl_2
hν

11.29 Mechanism of Benzene Side-chain Halogenation

The mechanism for side-chain chlorination is shown below.

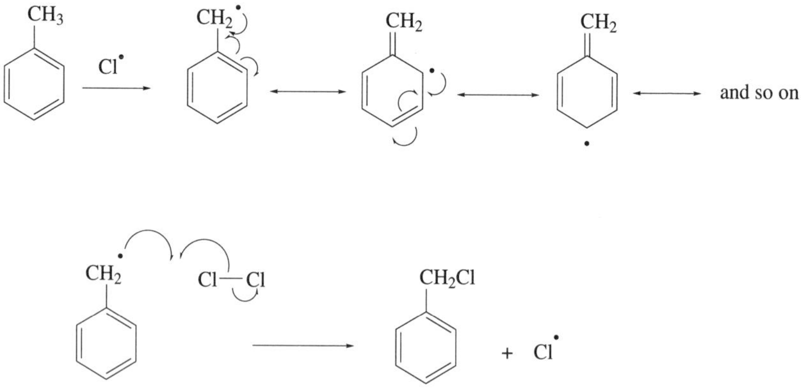

CH_3 $CH_2^{\bullet}$ CH_2 CH_2
$Cl^{\bullet}$
and so on

$CH_2^{\bullet}$ Cl—Cl CH_2Cl
+ $Cl^{\bullet}$

11.30 Effect of Alkali/Water on $PhCH_2Cl$, $PhCHCl_2$ and $PhCCl_3$

The halogenated side-chains of benzenes can be converted to alcohols, aldehydes or acids dependent on the number of halogens in the side-chain. The reaction is carried out using an aqueous alkali, or just water.

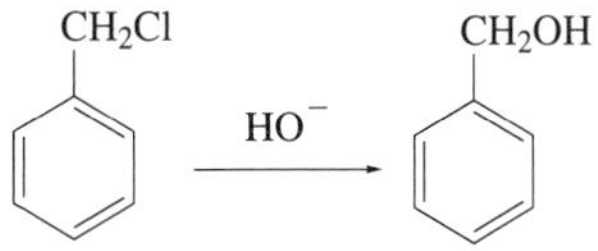

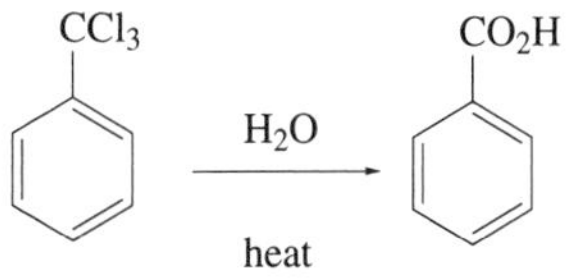

CH_2Cl $\xrightarrow{HO^-}$ CH_2OH

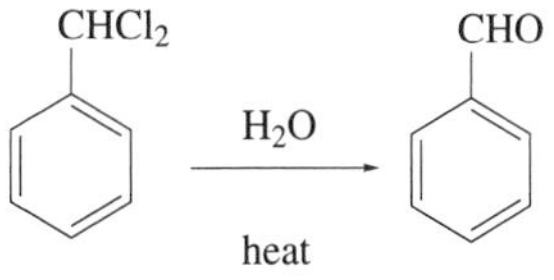

$CHCl_2$ $\xrightarrow[\text{heat}]{H_2O}$ CHO

CCl_3 $\xrightarrow[\text{heat}]{H_2O}$ CO_2H

11.31 Stabilised Carbocations of Benzene Systems

There are some carbocations which are stable because of the resonance stabilisation gained from being adjacent to benzene rings. One such system is shown below.

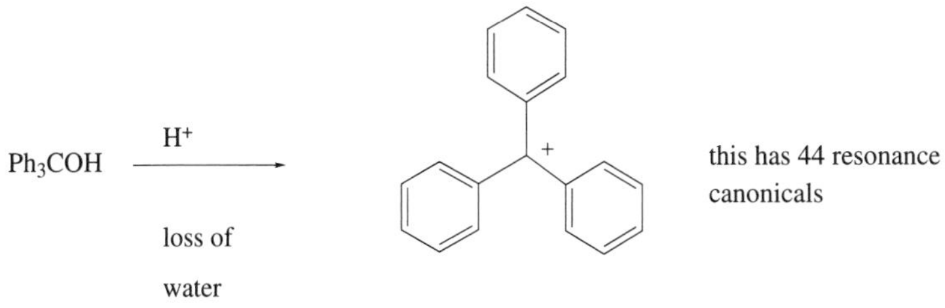

Ph_3COH $\xrightarrow[\text{loss of water}]{H^+}$ +

this has 44 resonance canonicals

11.32 Stabilised Carbanions of Benzene Systems

Benzylic carbanions are also stabilised, especially so when the ring contains -I/-R groups *o-/p-* to the benzylic position.

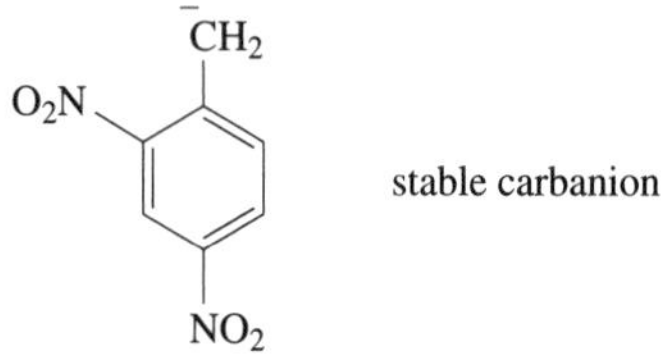

$\bar{C}H_2$ O_2N NO_2 stable carbanion

11.33 Conversion of Alkyl Side-chains on Benzene to Acids

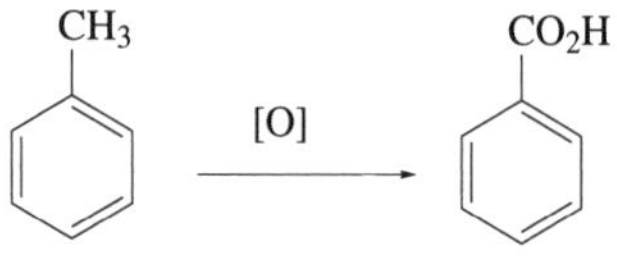

CH_3 [O] CO_2H

11.34 Conversion of Benzene Diacids to Anhydrides

This is also a chemical method of separation of regioisomers.

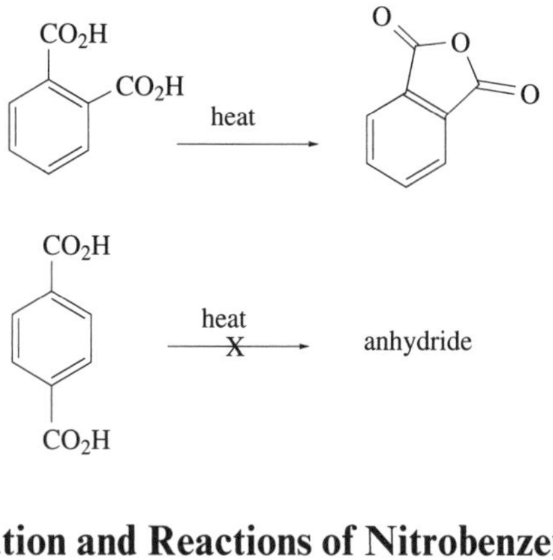

CO_2H CO_2H heat O O O

CO_2H CO_2H heat X anhydride

11.35 Nitration and Reactions of Nitrobenzenes

The reaction sequence is very important in producing the desired benzene derivative.

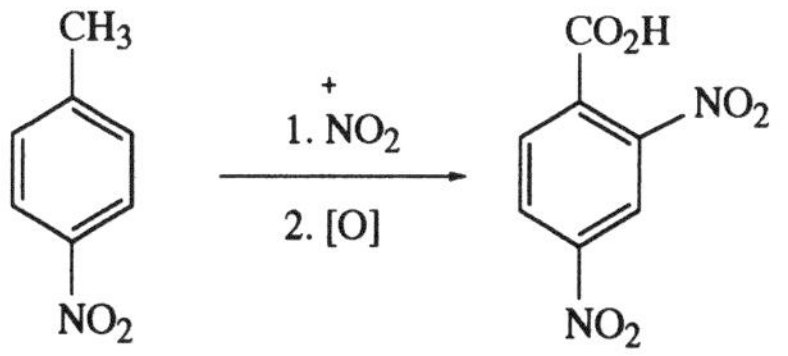

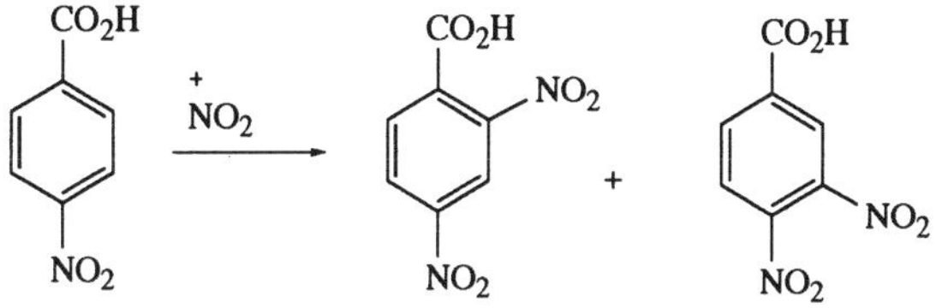

CH3 → 1. $\overset{+}{NO_2}$ 2. [O] → CO2H, NO2, NO2

CO2H, NO2 → $\overset{+}{NO_2}$ → CO2H, NO2, NO2 + CO2H, NO2, NO2

11.36 Conversion of Nitrobenzenes to Aminobenzenes: Selective Reduction Using Na_2S

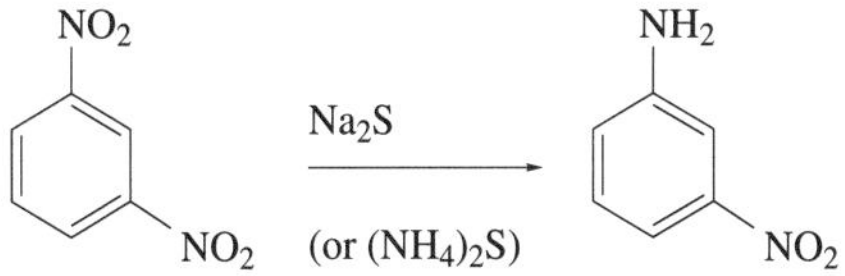

NO_2, NO_2 → Na_2S (or $(NH_4)_2S$) → NH_2, NO_2

The *m*-dinitrobenzene can be formed by controlled nitration of benzene.

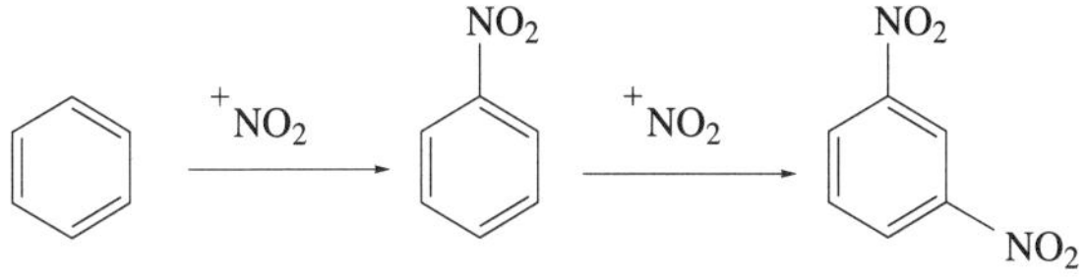

→ $\overset{+}{NO_2}$ → NO_2 → $\overset{+}{NO_2}$ → NO_2, NO_2

11.37 Charge Transfer Complexes of Benzenes

If there is a significant electron density difference between two substituted benzene rings, they will attract one another, as in the complex shown below. The nitro groups pull electron density out of the ring system, making it electron-deficient. The methyls push electron density into the ring system making it electron-rich.

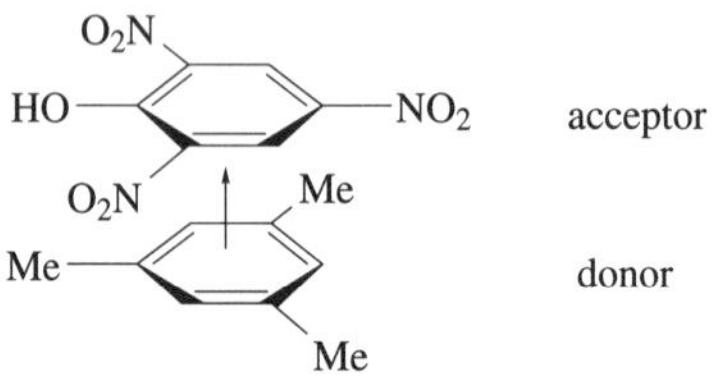

O_2N HO NO_2 acceptor
O_2N Me
Me donor
Me

crystalline, highly coloured

11.38 Conversion of Aminobenzenes to Nitrobenzenes

$$ArNH_2 \xrightarrow[Cu^+]{NaNO_2} ArNO_2$$ *via* diazonium salt

11.39 Conversion of Aminobenzenes to Amides

This is an important reaction as it changes directing properties of nitrogen. Note that the last step here is an oxidative conversion of **amino** to **nitro**.

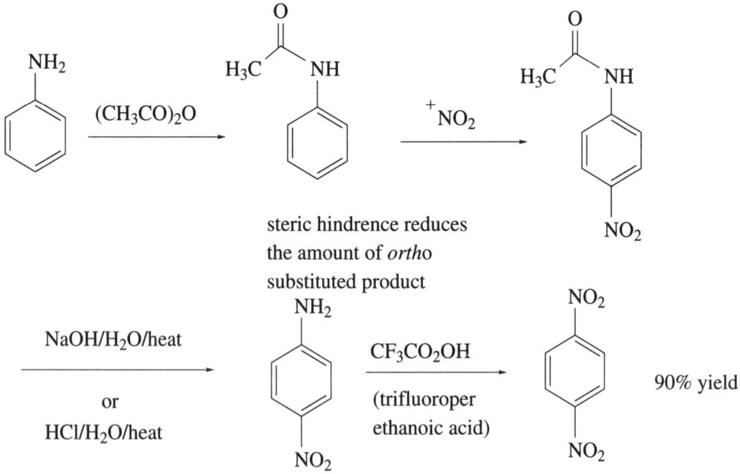

NH_2 $\xrightarrow{(CH_3CO)_2O}$ H_3C (C=O) NH $\xrightarrow{{}^+NO_2}$ H_3C (C=O) NH ... NO_2

steric hindrence reduces the amount of *ortho* substituted product

$\xrightarrow{NaOH/H_2O/heat}$ or HCl/H_2O/heat NH_2 ... NO_2 $\xrightarrow{CF_3CO_2OH}$ (trifluoroper ethanoic acid) NO_2 ... NO_2 90% yield

11.40 Conversion of Benzene Sulfonic Acids to Nitrobenzenes

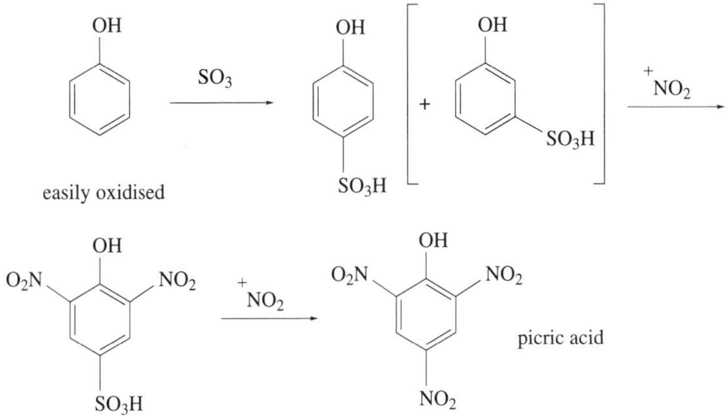

OH

SO_3

OH

OH

+

SO_3H

$\overset{+}{N}O_2$

easily oxidised

SO_3H

OH

O_2N NO_2

$\overset{+}{N}O_2$

OH

O_2N NO_2

picric acid

SO_3H

NO_2

11.41 Synthesis of Aniline, $PhNH_2$, and Substituted Anilines by Reduction of Nitro Compound

$$ArNO_2 \xrightarrow{\text{reduction}} ArNH_2$$

11.42 Anilines by Amination of Suitable Activated Aryl Halides

$$ArCl \longrightarrow ArNH_2$$

11.43 Anilines from Rearrangements

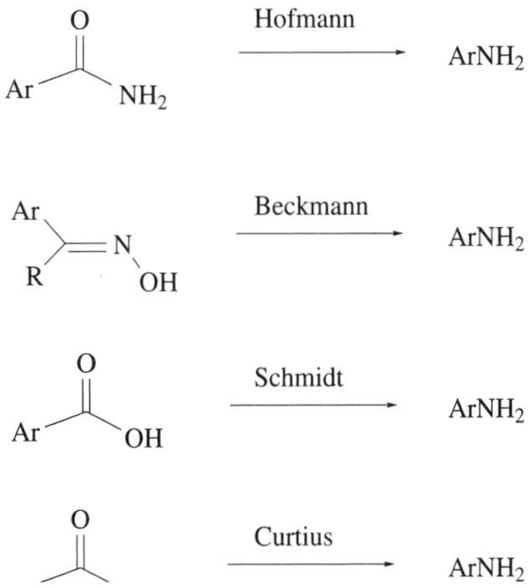

O

Ar NH_2

Hofmann

$ArNH_2$

Ar

R N OH

Beckmann

$ArNH_2$

O

Ar OH

Schmidt

$ArNH_2$

O

Curtius

$ArNH_2$

CH_3 CH_3 [O] loss of water → (phthalic anhydride) NH_3 heat/pressure → (phthalimide)

H^+/H_2O → CO_2H $CONH_2$ Hofmann → CO_2H NH_2

anthranilic acid

11.44 Properties of Aminobenzene

Aminobenzene is a weaker base than amino alkyls. The protons on the nitrogen of an *N*-substituted aminobenzene can be considered acidic, *i.e.* they can be removed by a strong base.

Ph—N(Ph)H —K→ Ph—N(Ph)$^-$ K^+ + R—X —RX, S_N2→ Ph—N(Ph)R

Aminobenzene undergoes electrophilic aromatic substitution. It is a highly activated system (conversion to amide might be necessary); NH_2 is *o*-/*p*- directing.

11.45 The Basicity of Aminobenzene

Aminobenzene is a very weak base, the pK_b is 9.4 (or the pK_a of the conjugate acid (p$K_{bH}{}^+$) is 4.6). The free amine has a possibility of forming resonance canonicals. Once the amino group is protonated it can no longer form these canonicals. The lone pair of the amino group in aminobenzene is not so available for donation. Aminobenzene is more stable as the free amine. Cyclohexylamine cannot form resonance canonicals and hence the lone pair is ready for donation. It has a pK_b of around 3.4 (or the pK_a of the conjugate acid (p$K_{bH}{}^+$) is 10.6). A compound such as the 2,4,6-trinitro-substituted aminobenzene below, has a

$pK_{bH}{}^{+}$ of -10; it does not like protons! The *N*,*N*-dimethyl-substituted analogue is 10000 times a more powerful base; why? Steric hindrance prevents the donation of the lone pair into the ring; the NMe_2 group must rotate out of the plane of the benzene ring to minimise the steric interactions. This is a stereoelectronic effect.

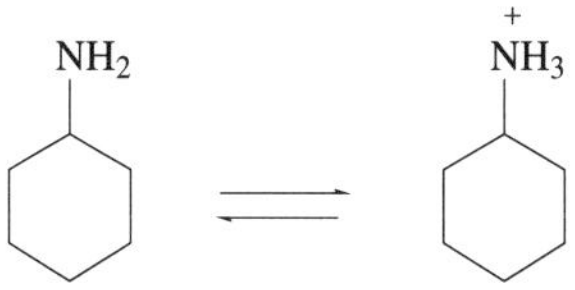

NH_2 ⇌ $\overset{+}{N}H_3$

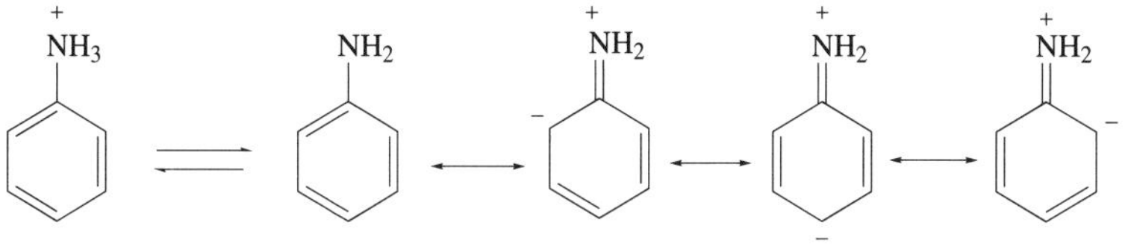

$\overset{+}{N}H_3$ ⇌ NH_2 ⟷ $\overset{+}{N}H_2$ ⟷ $\overset{+}{N}H_2$ ⟷ $\overset{+}{N}H_2$

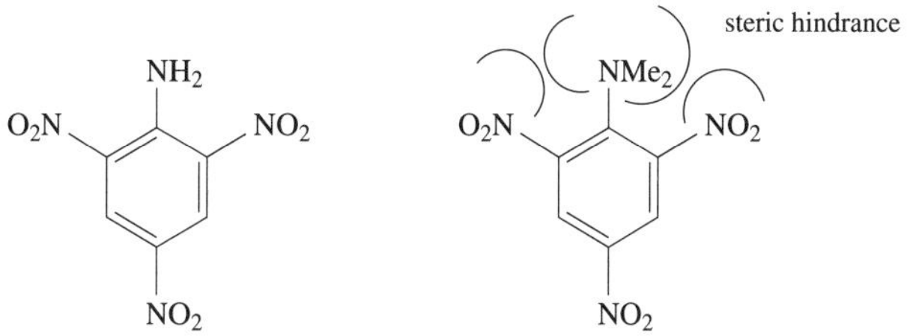

steric hindrance

NH_2, O_2N, NO_2, NO_2 — NMe_2, O_2N, NO_2, NO_2

11.46 Reactions of Aminobenzene: Oxidation of Aminobenzene

$$ArNH_2 \longrightarrow ArNO_2$$

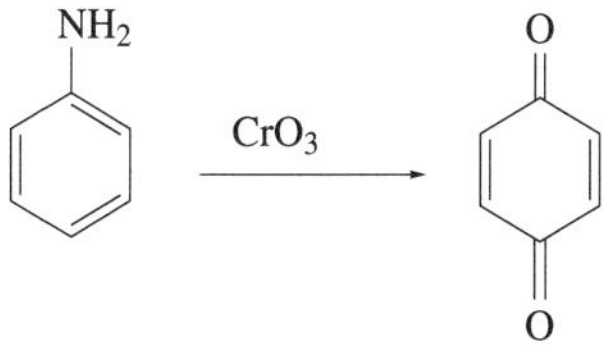

NH_2 $\xrightarrow{CrO_3}$ O, O

11.47 Reactions of Aminobenzene: Diazotation

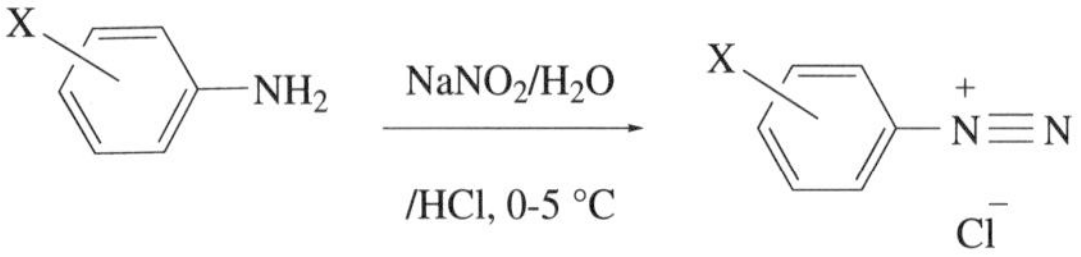

X-C$_6$H$_4$-NH$_2$ —($NaNO_2/H_2O$ /HCl, 0-5 °C)→ X-C$_6$H$_4$-$\overset{+}{N}\equiv N$ Cl^-

(this means substituent X on an undefined carbon of the ring)

aryl diazonium salt

11.48 Reactions of Secondary Aminobenzenes

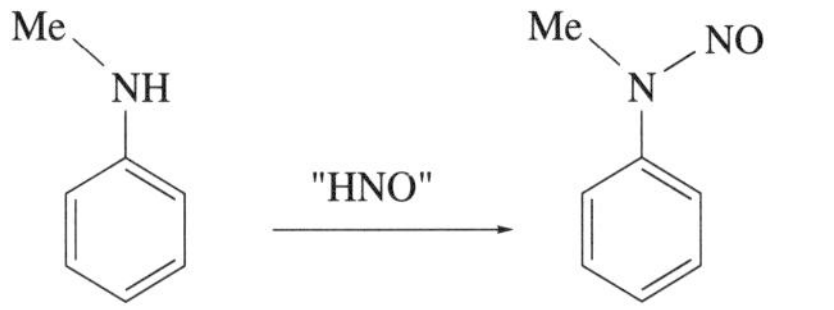

PhNHMe —("HNO")→ PhN(Me)NO

N-nitrosoamine

PhNH$_2$ → PhNHC(O)Me —(1. Na; 2. MeI)→ PhN(Me)C(O)Me —(HO^-)→ PhNHMe

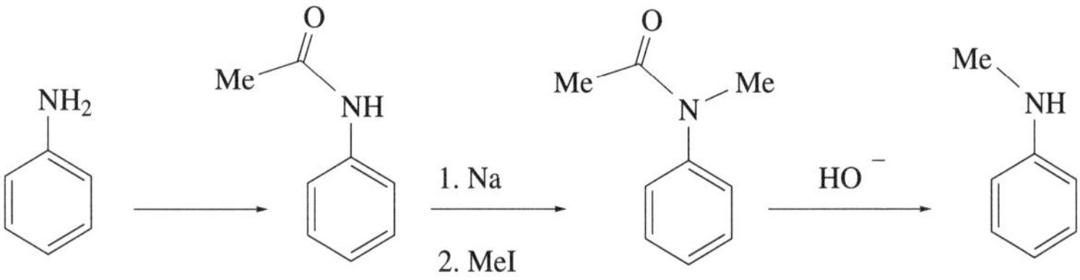

11.49 Reactions of Tertiary Aminobenzenes

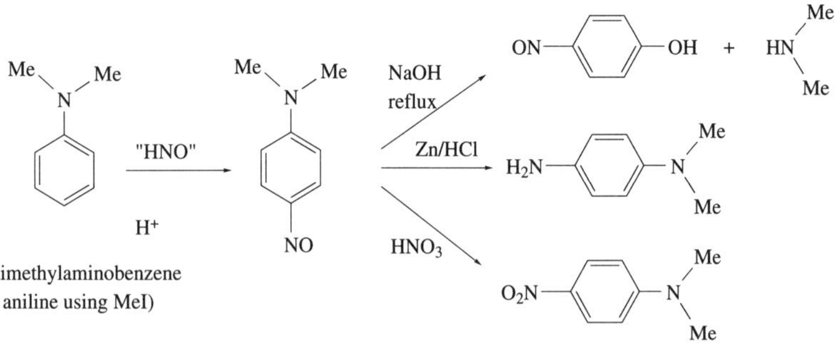

PhNMe$_2$ —("HNO", H^+)→ 4-ON-C$_6$H$_4$-NMe$_2$

N,N-dimethylaminobenzene (from aniline using MeI)

4-ON-C$_6$H$_4$-NMe$_2$ —(NaOH reflux)→ ON-C$_6$H$_4$-OH + HNMe$_2$

4-ON-C$_6$H$_4$-NMe$_2$ —(Zn/HCl)→ H$_2$N-C$_6$H$_4$-NMe$_2$

4-ON-C$_6$H$_4$-NMe$_2$ —(HNO_3)→ O$_2$N-C$_6$H$_4$-NMe$_2$

11.50 Reactions of Aminobenzene: Diazonium Salts

When forming diazonium salts, the anion of a strong acid is needed, Cl^- or BF_4^-. If the anion of a weak acid is used, then we obtain covalently bonded products, for example CN^-.

$$Ar{-}N{=}N{-}CN$$

11.51 Hydrolysis of Diazonium Salts

$$Ar\overset{+}{N}_2 \xrightarrow[\text{reflux}]{H_2O} ArOH + NH_2$$

11.52 The Sandmeyer Reaction

$$\text{2-}CH_3C_6H_4NH_2 \xrightarrow[H^+/HCl]{NaNO_2} \text{2-}CH_3C_6H_4\overset{+}{N}_2 \xrightarrow[\text{(or CuBr)}]{CuCl} \text{2-}CH_3C_6H_4Cl\ (Br)$$

$$Ar\overset{+}{N}_2 \xrightarrow{KI} ArI$$

$$Ar\overset{+}{N}_2 \xrightarrow{CuCN} ArCN \longrightarrow ArCOOH,\ ArCONH_2\ \textit{etc.}$$

11.53 The Schiemann Reaction

$$Ar\overset{+}{N}_2\ \overset{-}{Cl} \xrightarrow{HBF_4} Ar\overset{+}{N}_2\ \overset{-}{BF_4}$$

$$ArNH_2 \xrightarrow[H_2O/HBF_3]{NaNO_2} Ar\overset{+}{N}_2\ \overset{-}{BF_4}\ \text{(a crystalline salt)}$$

$$Ar\overset{+}{N}_2\ \overset{-}{BF_4} \xrightarrow{\text{heat}} ArF + N_2 + BF_3$$

$$Ar\overset{+}{N}_2\ \overset{-}{BF_4} \xrightarrow{NO_2/Cu} ArNO_2$$

11.54 Reduction of Diazonium Salts with H_3PO_2

$$ArN_2^+ \xrightarrow[H_2O]{H_3PO_2} ArH + H_3PO_3 + N_2$$

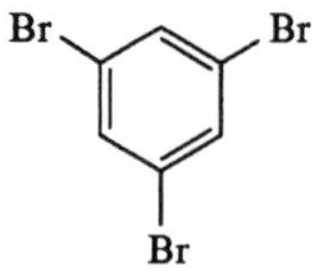

Br Br Br

this is not possible to make by direct bromination

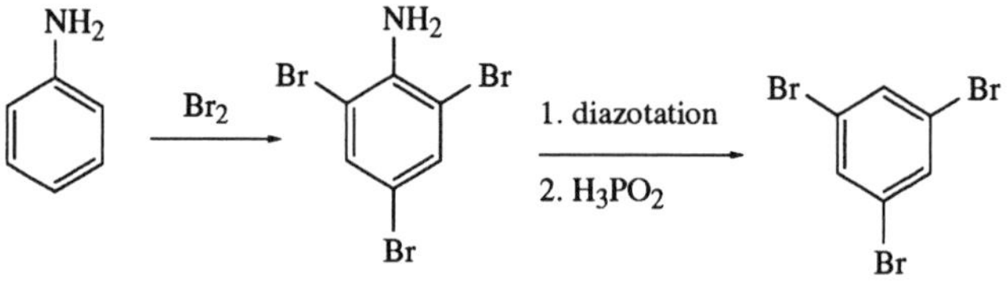

NH_2 Br_2 NH_2 Br Br Br 1. diazotation 2. H_3PO_2 Br Br Br

11.55 Generation of Benzyne from Diazonium Salts

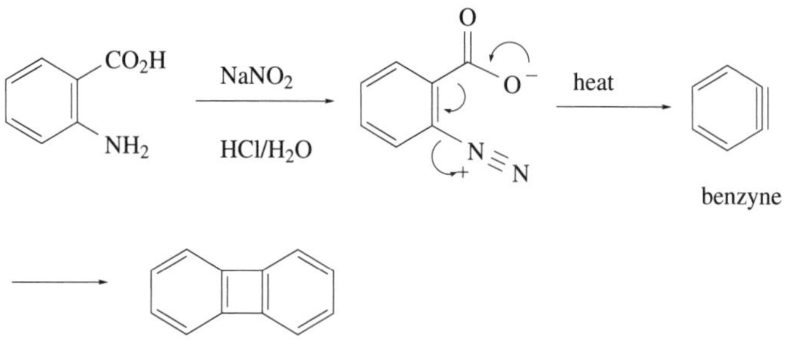

CO_2H NH_2 $NaNO_2$ HCl/H_2O O O⁻ N≡N heat

benzyne

11.56 Coupling of Diazonium Salts

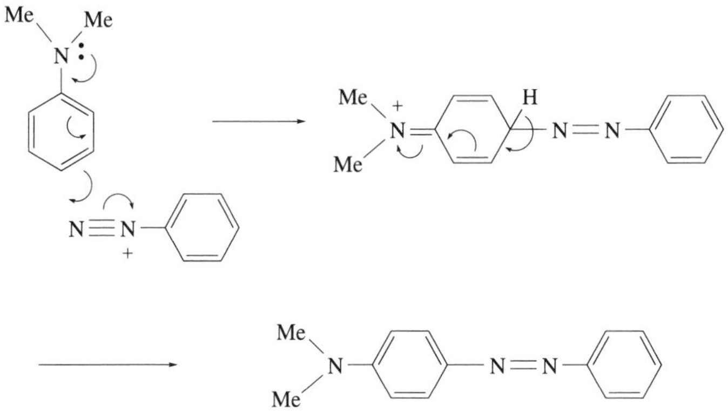

azo dye

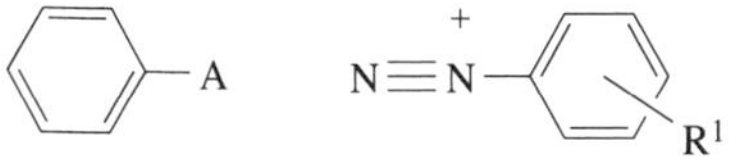

A = activator, (+**R**)

11.57 Reduction of Diazonium Salts to Aryl Hydrazines

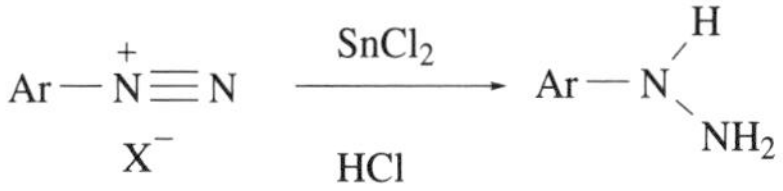

$$Ar-\overset{+}{N}\equiv N \; X^- \xrightarrow[HCl]{SnCl_2} Ar-N(H)NH_2$$

aryl hydrazide

11.58 Aryl Halides

The dipole moments are smaller and C–halogen bond lengths are shorter for aryl halides; therefore they are much less willing to be broken.

11.59 Synthesis of Aryl Halides

There are two general approaches to the synthesis of aryl halides: these are direct halogenation and from amines *via* the Sandmeyer or Schieman reactions.

$$m\text{-}C_6H_4(NO_2)_2 \xrightarrow{Sn/HCl} m\text{-}C_6H_4(NH_2)_2 \xrightarrow[2.\ CuCl]{1.\ diazotize} m\text{-}C_6H_4Cl_2$$

$$m\text{-}C_6H_4(NO_2)_2 \xrightarrow{(NH_4)_2S} m\text{-}H_2NC_6H_4NO_2 \xrightarrow[2.\ CuCl]{1.\ diazotize} m\text{-}ClC_6H_4NO_2 \xrightarrow[2.\ diazotize;\ 3.\ CuBr]{1.\ reduce} m\text{-}ClC_6H_4Br$$

11.60 Reactions of Aryl Halides: Formation and Coupling of Grignards

$$C_6H_5Br \xrightarrow{Mg} C_6H_5MgBr \quad \text{reacts like an alkyl Grignard}$$

$$C_6H_5Cl \xrightarrow{Li} C_6H_5Li \quad \text{reacts like an alkyl lithium}$$

$$ArI \xrightarrow{Cu/heat} Ar{-}Ar$$

11.61 Nucleophilic Displacement of Activated Aryl Halides

See 11.66.

11.62 The Preparation of Sulfonic Acids

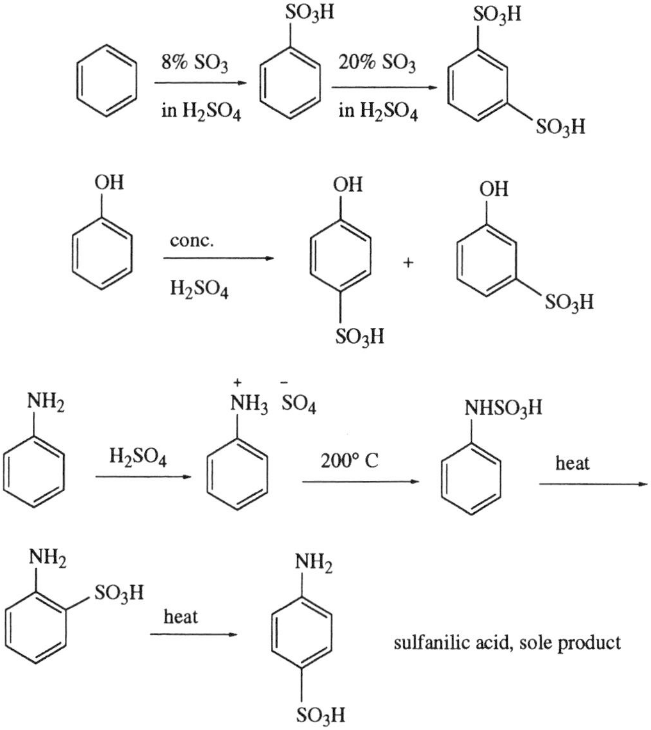

SO_3H SO_3H

8% SO_3 in H_2SO_4 → 20% SO_3 in H_2SO_4 →

SO_3H

OH OH OH

conc. H_2SO_4 → +

SO_3H

SO_3H

NH_2 $\overset{+}{N}H_3$ $\overset{-}{S}O_4$ $NHSO_3H$

H_2SO_4 → 200° C → heat →

NH_2 NH_2

SO_3H

heat →

sulfanilic acid, sole product

SO_3H

11.63 Reactions of Sulfonic Acids

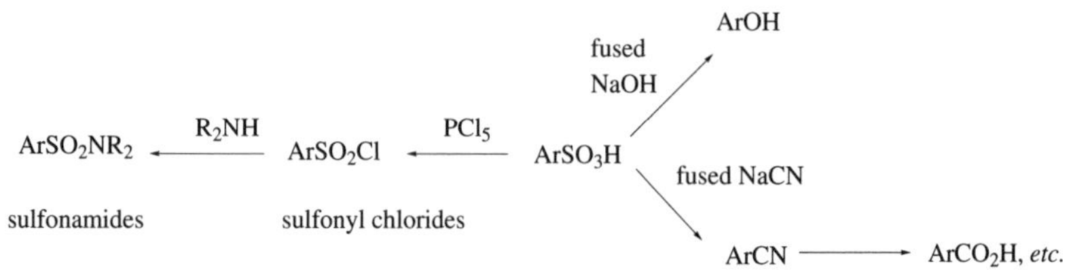

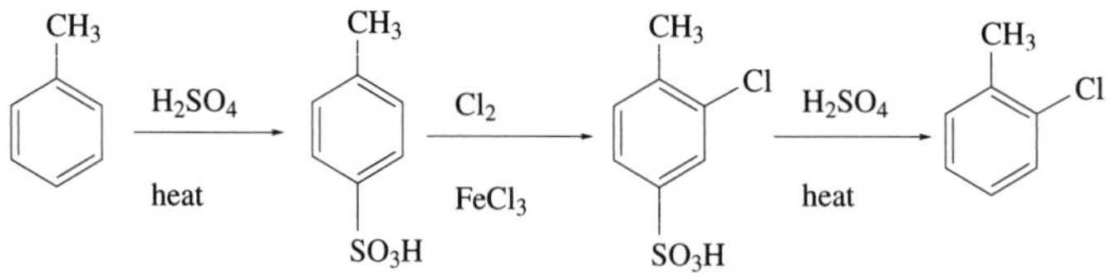

ArOH
fused NaOH
$ArSO_2NR_2$ ← (R_2NH) $ArSO_2Cl$ ← (PCl_5) $ArSO_3H$
sulfonamides
sulfonyl chlorides
fused NaCN
ArCN → $ArCO_2H$, *etc.*

CH_3, H_2SO_4, heat, CH_3, SO_3H, Cl_2, $FeCl_3$, CH_3, Cl, SO_3H, H_2SO_4, heat, CH_3, Cl

11.64 Preparation and Reactions of Phenols

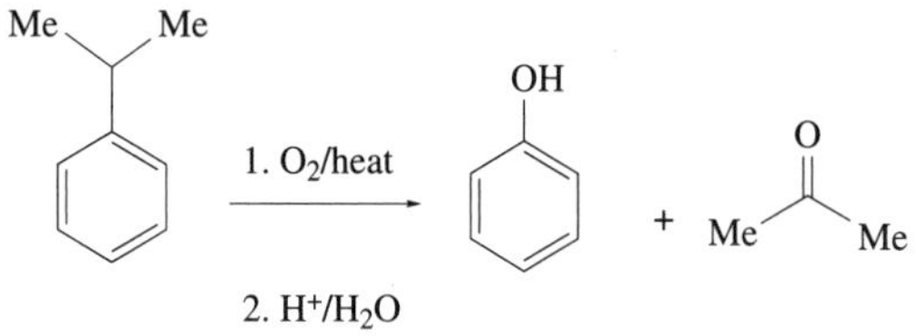

Me Me
1. O_2/heat
2. H^+/H_2O
OH
+ O, Me, Me

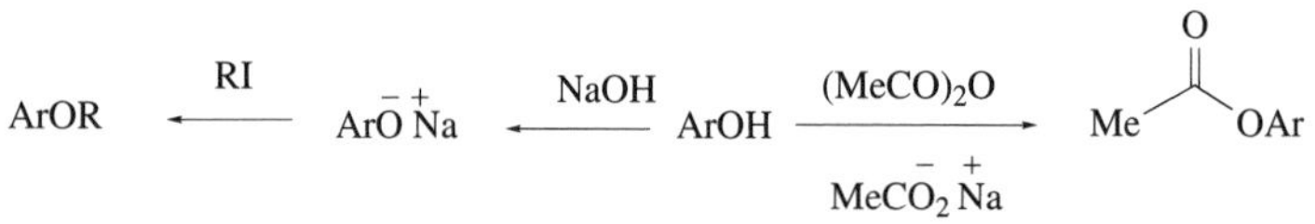

ArOR ← (RI) $Ar\bar{O}\overset{+}{N}a$ ← (NaOH) ArOH → ($(MeCO)_2O$, $MeCO_2^-\,Na^+$) O, Me, OAr

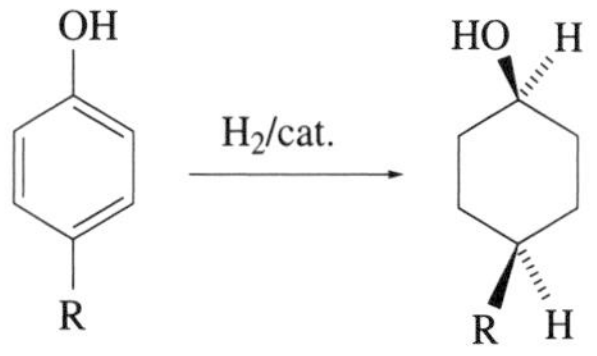

OH / R — H_2/cat. → HO H / R H *cis* derivative

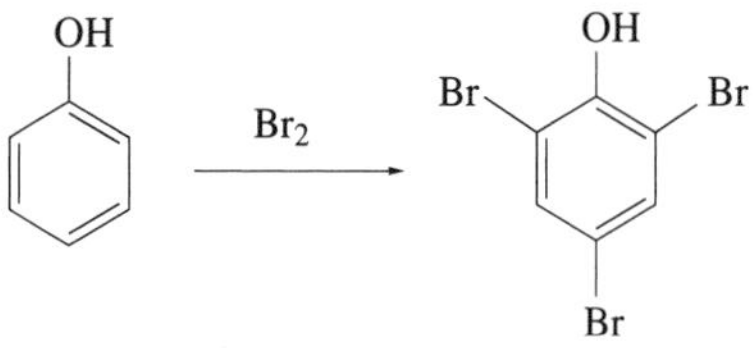

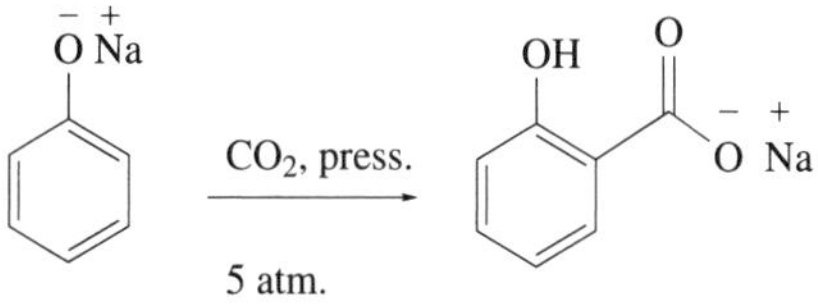

OH — Br_2 → OH, Br, Br, Br

$\overset{-}{O}\overset{+}{Na}$ — CO_2, press. / 5 atm. → OH, O, $\overset{-}{O}\ \overset{+}{Na}$

ortho substitution

11.65 Acidity of Phenols

The pK_a of cyclohexanol is 16: it is a weak acid. The pK_a of phenol is 10: a relatively strong acid. Why should there be such a large difference?

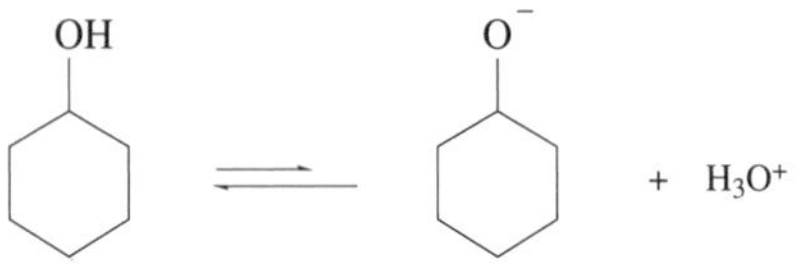

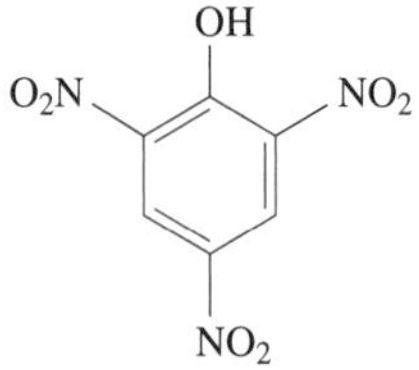

OH ⇌ O^- + H_3O^+

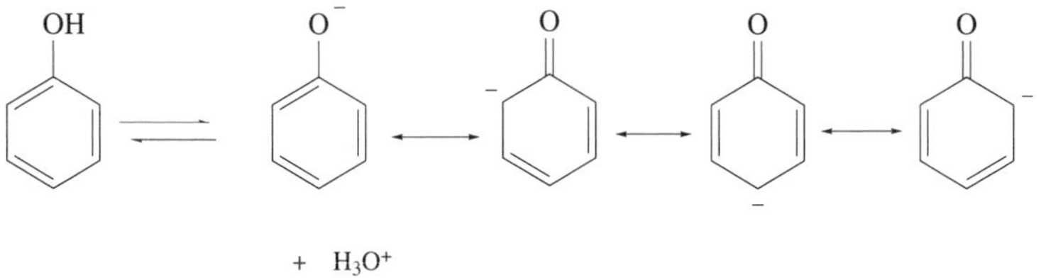

OH ⇌ O^- ⟷ O ⟷ O ⟷ O

+ H_3O^+

OH

O_2N NO_2

NO_2

picric acid

The anion of cyclohexanol has no resonance canonicals and is therefore much less stable than the anion of phenol which has many resonance canonicals. If substituents are in the benzene ring which produce further delocalisation (more resonance canonicals), then the acidity of the hydroxy group increases; these substituents are the **-I**, **-R** type. For example, *o*-nitrophenol has a pK_a of 7.24, and *p*-nitrophenol of 7.14. The *m*-nitrophenol can only use its **-I** effect and hence has a pK_a of only 8.35. The trinitro-substituted picric acid has a pK_a of 1.02: this is a very strong acid.

11.66 Benzenes and Nucleophilic Aromatic Substitution (NAS)

For nucleophilic attack on a benzene ring to take place a good anionic leaving group would be required. Even then, direct nucleophilic displacement does not occur in halobenzenes under normal circumstances. NAS takes place by (a) an addition–elimination mechanism or (b) an elimination–addition mechanism.

11.67 NAS and the Addition–Elimination Mechanism

The intermediate is very unstable. It can be stabilised by delocalisation of the negative charge by means of **-I** or **-I/-R** groups. These groups will be most effective when *o-/p-* to the leaving group because these are the positions at which the negative charge resides. This type of group **activates** the ring **WITH RESPECT TO NAS NOT EAS – DO NOT CONFUSE THIS.**

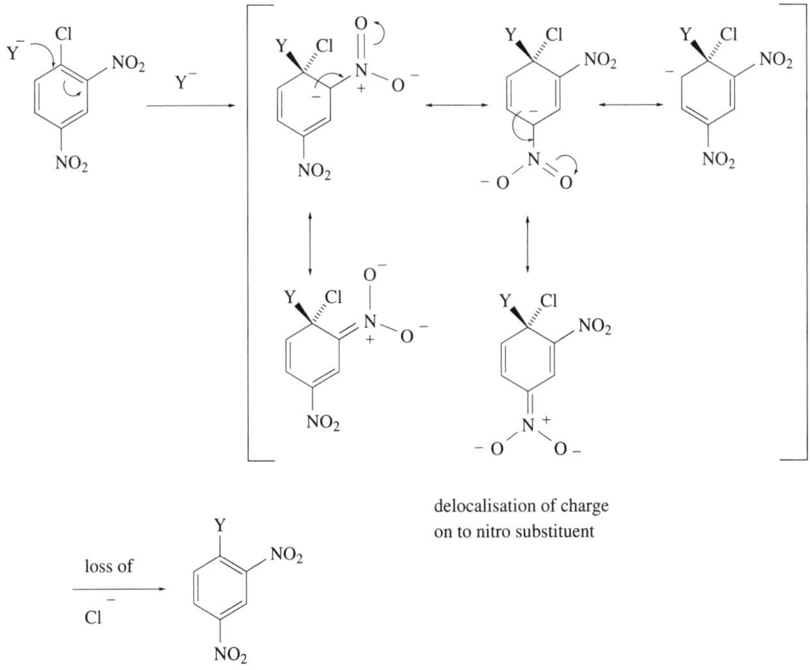

Y⁻ Cl NO2 NO2 Y⁻ Y Cl O N O⁻ NO2 NO2 Y Cl NO2 N O O NO2 NO2 Y Cl O⁻ N O⁻ NO2 Y Cl NO2 N O O

delocalisation of charge on to nitro substituent

loss of Cl⁻ Y NO2 NO2

SUBSTITUENTS WHICH ENCOURAGE EAS DISCOURAGE NAS AND *VICE VERSA*.

11.68 NAS and the Elimination–Addition Mechanism

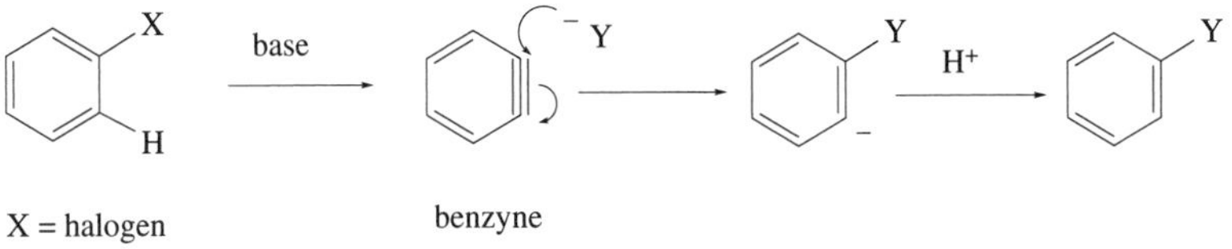

X H base Y⁻ Y Y H⁺ Y

X = halogen

benzyne

Some examples of this type of reaction are given below.

F PhLi Ph

OMe $NaNH_2$, NH_3 OMe

Br H_2N

11.69 Relevant Questions

12(b), 15, 17, 20, 25, 29, 34, 54(iv), 60, 61, 72, 78, 82, 83, 91, 98, 100(b), 106(c), 107, 109, 112, 113, 114, 120, 121, 122, 127, 136, 141, 143(b), 160(a), 161(b), 163, 166, plus parts of functional group interconversion questions.

APPENDIX

Exam Questions

This appendix contains exam questions collected from universities in the British Isles. On average the candidate is required to answer four questions in three hours; however, this can vary from six questions in three hours to four questions in two hours.

Many of the questions in this appendix have not been listed at the end of a given chapter. This is because they are functional group interconversion questions and may require knowledge of several chapters. Examples of these questions are 37, 56, 90, 103, *etc.* All functional group interconversion questions should be attempted.

The questions are from the end-of-year examinations of: Loughborough University of Technology; Dublin City University; University College Dublin; University of Warwick; University of Southampton; University of Wales, College of Cardiff; University of Dublin, Trinity College; The Queen's University of Belfast (questions 76, 77, 86, 93–98, 116–120, 128–130); and others that have requested to remain anonymous.

You should attempt these questions without reference to your notes and you should not need more than thirty-five to forty minutes per question.

1. Compare and contrast the bonding in ethane, ethene (ethylene) and ethyne (acetylene), with particular reference to the relationship between hybridisation and molecular shape.

   Indicate the hybridisation of the orbitals associated with the carbon atoms indicated by the arrow in each of the compounds shown below.

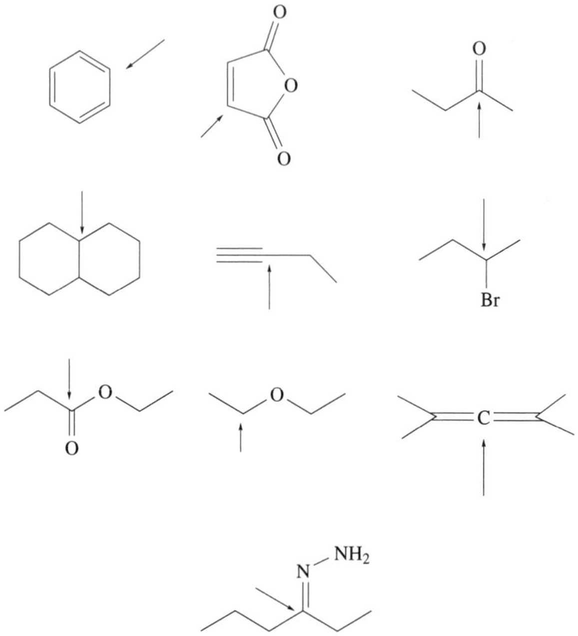

O O O O O Br O O O C NH_2 N

2. Predict the product of each of the following reactions:

(a) i. B_2H_6 ii. HO^-/H_2O_2

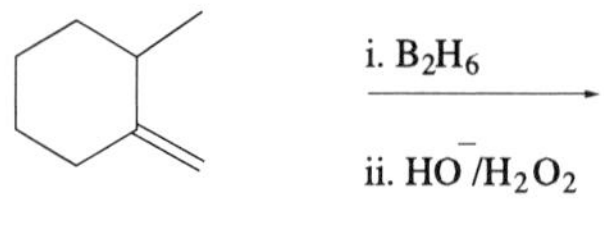

(b) Me H Br Me H HO i. K_2CO_3, acetone (solvent) ii. aq. NH_3, heat

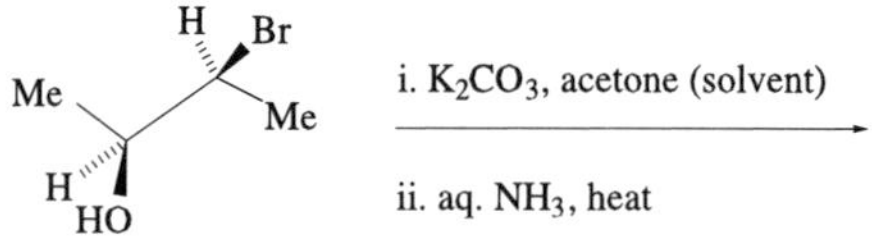

(c) Cl NaOEt/EtOH heat

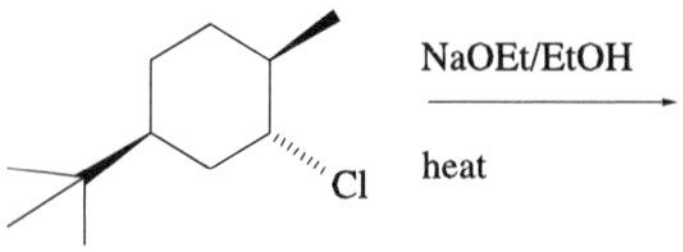

(d) CHO i. NH_3, HCN ii. H_3O^+, heat

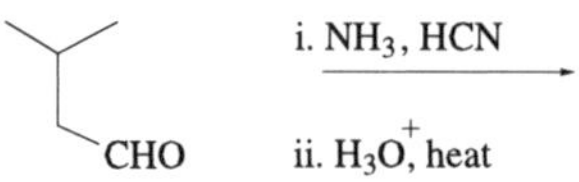

3. (a) Write notes on **TWO** methods of preparing alkenes.
   (b) Treatment of (1*R*, 3*S*)-1-chloro-3-methylcyclopentane with potassium *t*-butoxide in *t*-butyl alcohol gave two isomeric alkenes. Only one of the two products was optically active. Draw (1*R*, 3*S*)-1-chloro-3-methylcyclopentane indicating clearly its stereochemistry and write down the structures of the two product alkenes.
4. Write short notes on the following topics:
   (a) The use of ozonolysis in organic chemistry.
   (b) Radicals in organic chemistry.
   (c) Stereochemical features of elimination reactions.
5. "In some of their reactions, aldehydes and ketones can undergo nucleophilic attack; in other reactions, nucleophilic carbanions can be generated from them."

   Discuss this statement and include in your discussion detailed description of examples of the two classes of reaction.
6. Deduce the structures of the compounds **A** to **E** in the sequence below.

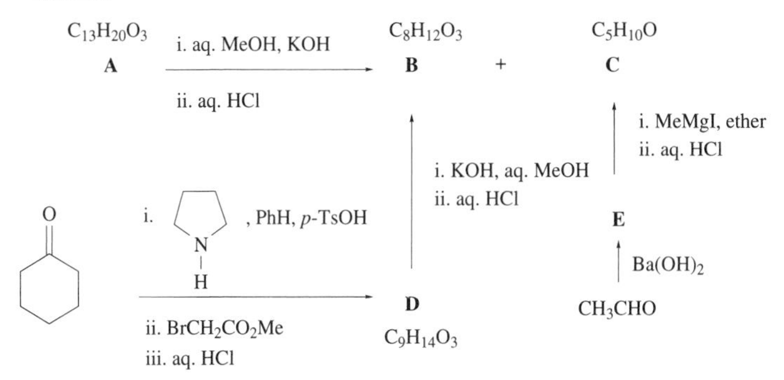

7. Suggest methods for carrying out **THREE** of the following transformations. Discuss briefly the mechanisms of the reactions you propose.

(a) $MeCH_2CHO \longrightarrow MeCH_2CH(OH)CO_2H$

(b) $MeCOCH_2CH_2CO_2Et \longrightarrow MeCOCH_2CH_2CH_2OH$

(c) $MeCOCH_2Me \longrightarrow Me_2C(OH)CH_2Me$

(d) $Me{-}C{\equiv}CH \longrightarrow Me{-}C{\equiv}C{-}CH_2Me$

8. Suggest mechanistic pathways that account for each of the following transformations.

(a) Proline (pyrrolidine-2-carboxylic acid, N–H, CO_2H) $\xrightarrow[\text{ii. } H_2O]{\text{i. } Ac_2O,\ HCO_2H}$ N-formylproline (N–CHO, CO_2H)

(b) $PhCO_2H$ $\xrightarrow[\text{ii. } Me_2NCH_2CH_2OH]{\text{i. } SOCl_2}$ $PhCO_2CH_2CH_2NMe_2$

(c) Cyclohexanone $\xrightarrow[\text{ii. NaH, } Et_2O \text{; iii. MeI}]{\text{i. } CH_3CHO,\ \text{aq. } Na_2CO_3}$ 2-(1-methoxyethyl)cyclohexanone (O, Me, OMe)

(d) $MeOCH_2CN$ $\xrightarrow[\text{ii. } H_3O^+]{\text{i. PhMgBr, } Et_2O}$ $PhCOCH_2OMe$ (O, OMe)

9. A chiral compound **A**, C_7H_{12}, on oxidation with hot concentrated chromic acid, gave two moles of acetic acid for every mole of **A** oxidised. On treatment with potassium permanganate under mildly basic conditions, compound **A** gave a product **B**, $C_7H_{14}O_2$. When compound **B** was treated with one molar equivalent of sodium periodate ($NaIO_4$), a product **C** was obtained which was readily oxidised to a carboxylic acid **D**, $C_7H_{12}O_2$. The acid **D**, on treatment with sodium hypoiodite (prepared by dissolving KI_3 in dilute sodium hydroxide solution) followed by acidification, gave a dicarboxylic acid **E**, $C_6H_{10}O_4$, which on treatment with methanol and a catalytic amount of concentrated sulfuric acid gave a neutral compound **F**, $C_8H_{14}O_4$. The acid **E** could not be resolved into optical isomers.

Draw the structures of compounds **A** to **F** and comment briefly on the reactions described.

10. Indicate how you would carry out the following transformations. State reagents and experimental conditions.

(a)

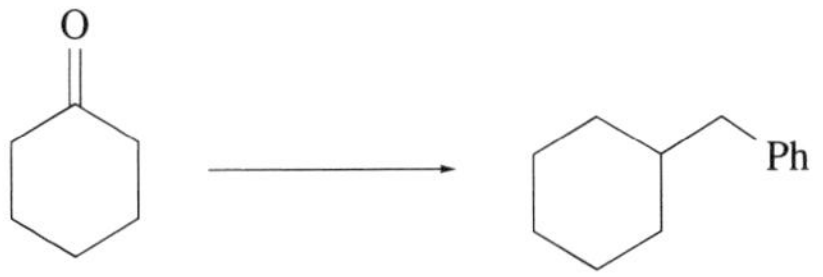

Ph

(b)

O

Ph

O

Ph

O

(c)

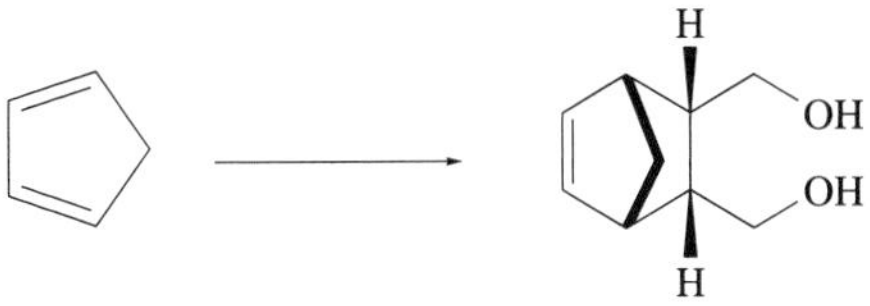

H

OH

OH

H

Me

NMe$_2$

(d)

Me

OH

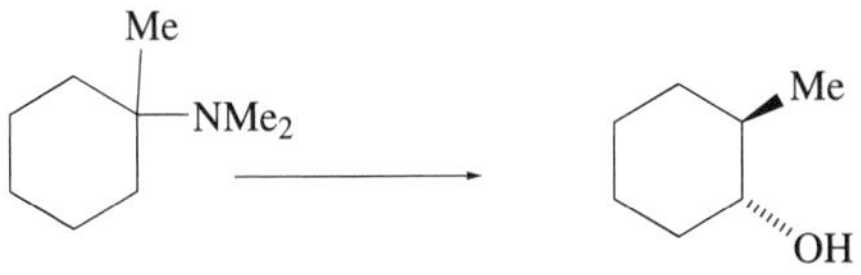

11. Explain briefly **EACH** of the following observations and comment on the influence of the solvent in each case.

(a) $C_2H_5CH(Br)CH_3$ + NaOH $\xrightarrow{\text{aqueous ethanol}}$ $C_2H_5CH(OH)CH_3$

optically active — optically active

(b) $C_6H_5CH(Cl)CH_3$ + NaOH $\xrightarrow{\text{aqueous acetone}}$ $C_6H_5CH(OH)CH_3$

optically active — optically inactive

12. (a) Alkynes are useful intermediates in organic chemistry. Discuss this statement and illustrate your answer with well-chosen examples.

(b) Describe how each of the following compounds could be made from the starting material given in brackets. Give reagents and conditions.

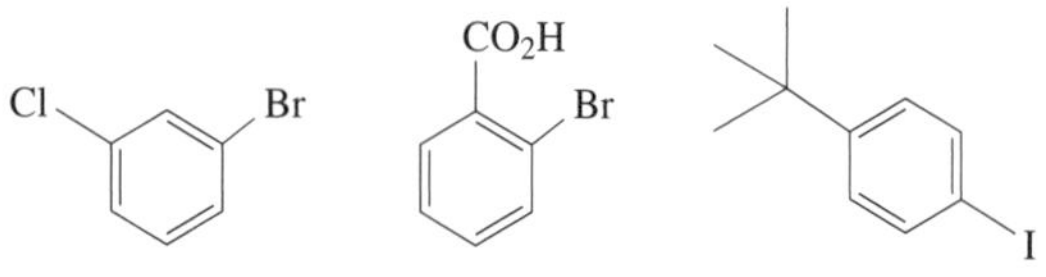

Cl, Br (benzene) — CO_2H, Br (benzene) — I (toluene)

(benzene) (benzene) (toluene)

13. Answer **TWO** of the following:

(a) Show how stereochemical studies have provided evidence for the S_N1 and S_N2 mechanisms.

(b) Describe the stereochemistry of the tartaric acids (HOOCCH(OH)CH(OH)COOH).

(c) Describe methods that can be used for the optical resolution of racemic mixtures.

(d) Describe classes of organic chiral compounds that do not contain central chirality (asymmetric carbon atoms).

14. Answer all the parts.

Predict the products of the following reactions. Indicate mechanisms using curly arrows. Given that the reactant is optically active, briefly explain in each case whether or not the product(s) will be optically active.

(a)  Ph, Me, H, Br — KCN / THF →

(b) 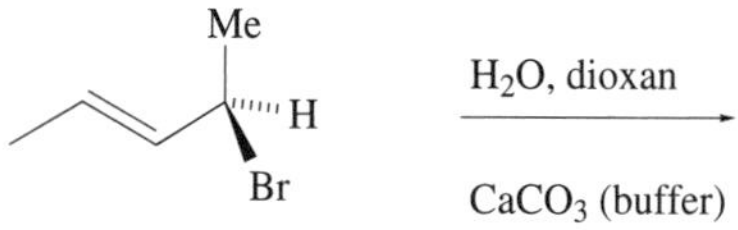 Me, H, Br — H_2O, dioxan / $CaCO_3$ (buffer) →

(c) 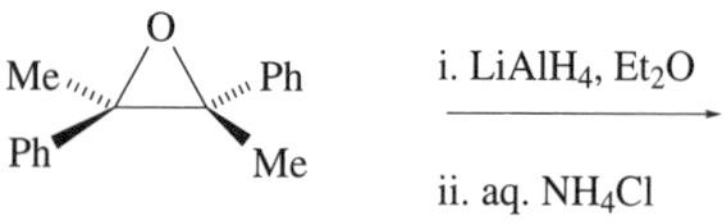 O, Me, Ph, Ph, Me — i. $LiAlH_4$, Et_2O / ii. aq. NH_4Cl →

(d) 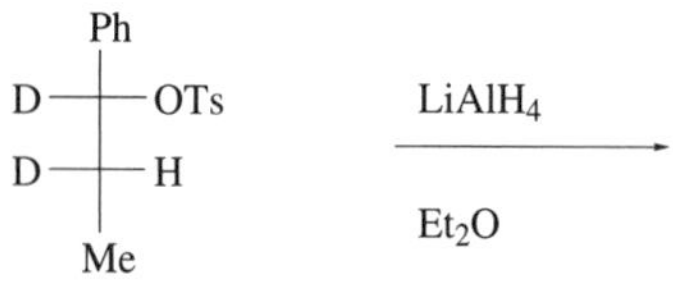 Ph, D, OTs, D, H, Me — $LiAlH_4$ / Et_2O →

Fischer projection

Give the absolute configuration of the product in (a) using the *R*/*S* nomenclature.

15. Write brief notes on the mechanisms of the two reactions which involve nucleophilic displacement at aryl carbon atoms.

Predict the products of **TWO** of the following reactions:

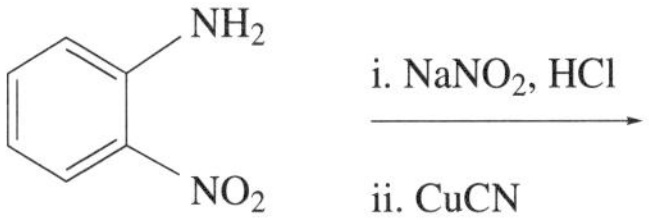

NH_2, NO_2 — i. $NaNO_2$, HCl; ii. CuCN →

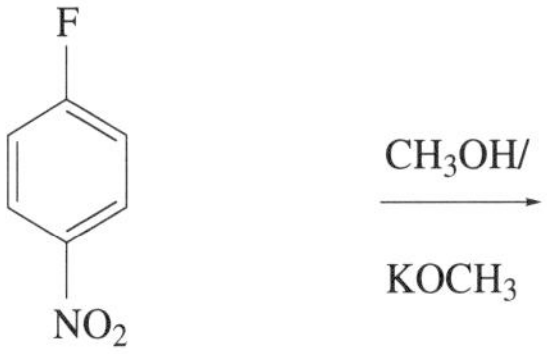

F, NO_2 — CH_3OH/$KOCH_3$ →

Cl, * — KNH_2, NH_3 →

* = [14]C labelled

16. Briefly explain the following observations.

(a) The bromide **A** reacts rapidly with ethanolic potassium cyanide, whereas the bromide **B** does not.

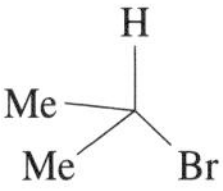

H, Me, Me, Br — **A**

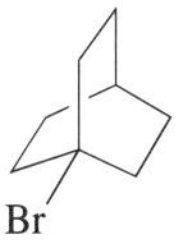

Br — **B**

(b) When an aqueous solution of **C** is treated with chromic acid, the acid **D** is formed.

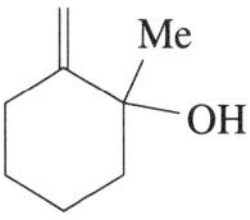

Me, OH — **C**

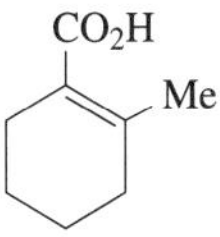

CO_2H, Me — **D**

(c) When **F** is heated a single olefinic product **E** is formed. When compound **G** is heated the same olefinic product **E** is also formed.

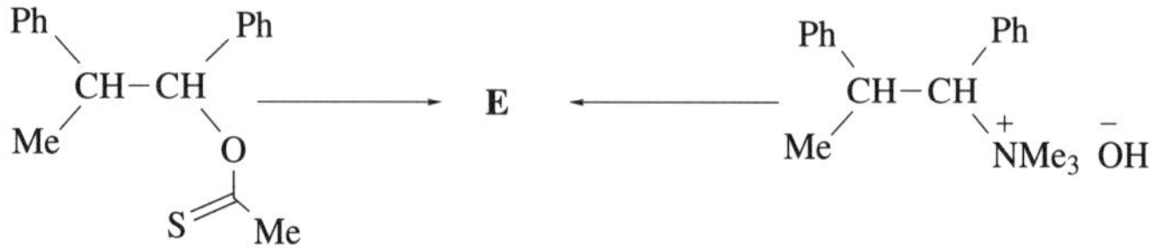

Ph–CH(Me)–CH(Ph)–O–C(=S)Me ⟶ **E** ⟵ Ph–CH(Me)–CH(Ph)–$\overset{+}{N}Me_3$ $\overset{-}{O}H$

F *erythro* **G** *threo*

(d) When ketone **H** is treated with $NaBH_4$, and the product is treated with acid chloride **I**, two different compounds are produced in equal amounts, which can be separated by chromatography.

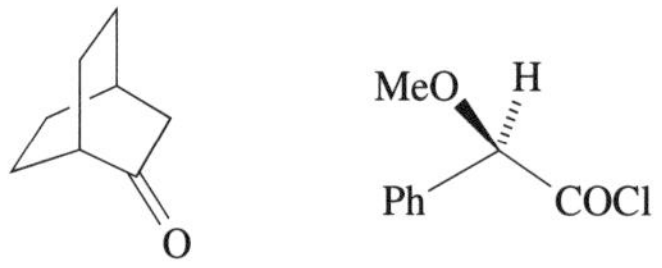

MeO H Ph COCl

H **I**

17. (i) Write down the major product of **EACH** of the following reactions:

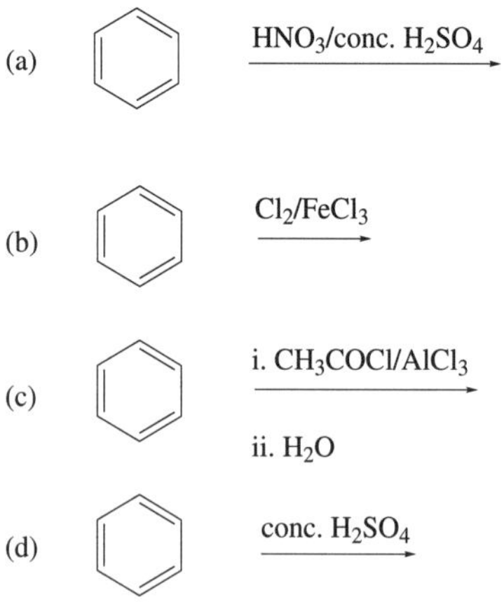

(a) benzene $\xrightarrow{HNO_3/\text{conc. } H_2SO_4}$

(b) benzene $\xrightarrow{Cl_2/FeCl_3}$

(c) benzene $\xrightarrow[\text{ii. } H_2O]{\text{i. } CH_3COCl/AlCl_3}$

(d) benzene $\xrightarrow{\text{conc. } H_2SO_4}$

(ii) Give detailed mechanisms, including the structures of intermediates, for **TWO** of the reactions in (i).

18. Stating reagents and conditions, indicate how you would carry out the following transformations. Highlight the important mechanistic, and where appropriate, stereochemical details.

(a)

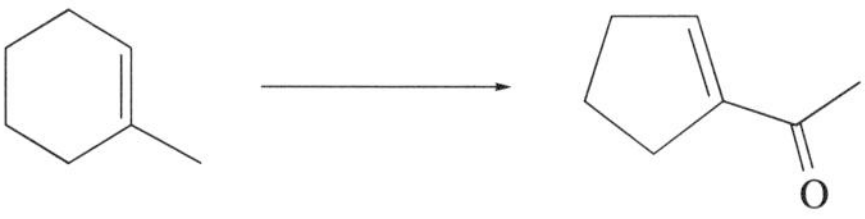

(b)

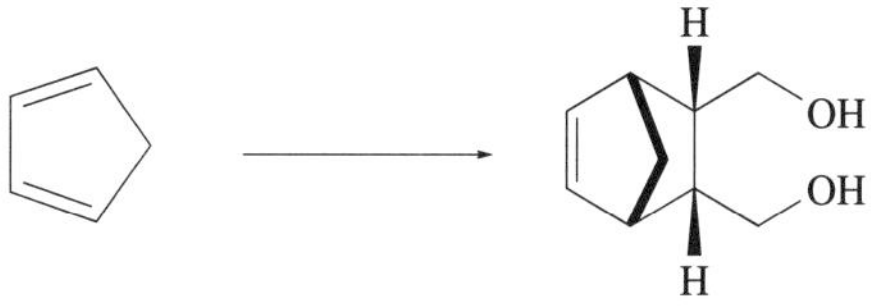

(c)

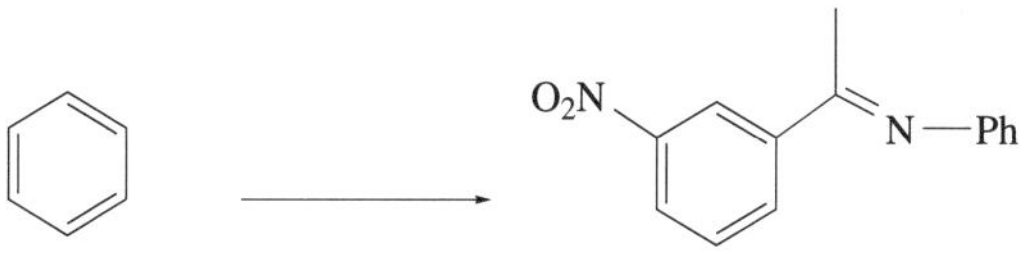

(d) $PhNH_2$ and $PhSO_3H$ ⟶ $PhN(SO_2Ph)(Me)$

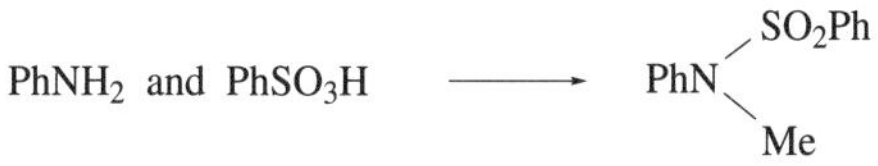

19. Write an essay on the hydrolysis of esters by acids and alkalis, with special reference to the mechanisms involved.

20. With well chosen examples discuss how inductive and resonance effects can influence electrophilic aromatic substitution.

Describe how the following molecules could be made from either benzene or toluene. Give reagents and conditions.

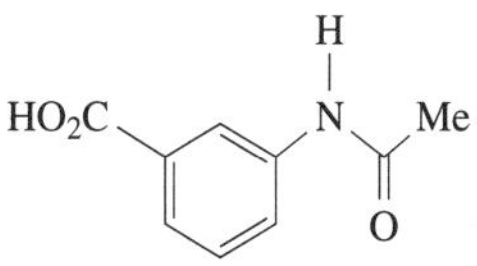

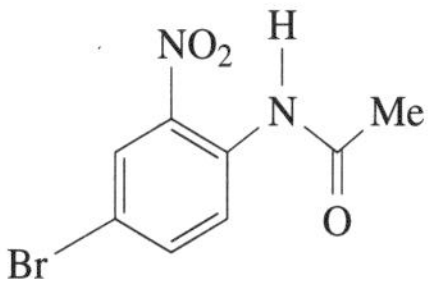

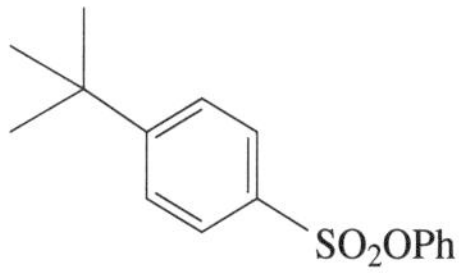

21. Outline **THREE** methods for the synthesis of primary amides and give possible reaction mechanisms for your synthetic routes.

One possible method for the synthesis of compound **A**, a precursor of nylon, is the hydrolysis of the cyclic amide **B**. Describe how you would prepare **B** from cyclohexanone and hydroxylamine (NH_2OH).

$NH_2(CH_2)_5COOH$

A **B**

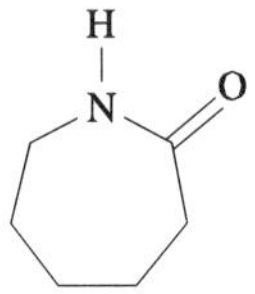

22. (i) With due attention to essential experimental detail, explain concisely how you would prepare an ethereal solution of methyl magnesium iodide from methyl iodide.

(ii) Predict the products from each of the following reactions.

(a) $PhCO_2Et$ → i. MeMgI, Et_2O; ii. H_3O^+

(b) Ph–(epoxide) → i. MeMgI, Et_2O; ii. H_3O^+

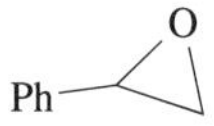

(c) Ph–≡–H → i. MeMgI, Et_2O; ii. CH_3CHO

(c) $PhCH_2CO_2H$ → i. Excess MeLi; ii. H_3O^+

23. Write brief notes on **THREE** of the following:

(a) The addition of hydrogen cyanide to aldehydes.
(b) The reduction of esters by lithium aluminium hydrides.
(c) The Beckmann rearrangement of oximes.
(d) The haloform reaction.
(e) The formation of hemiacetals and acetals.

24. (a) Discuss the conformational properties of mono- and disubstituted cyclohexane derivatives.

(b) Predict the relative thermodynamic stabilities of the two isomers for each of the following pairs of compounds. Give reasons for your predictions.

(a)

Me | | Me
MeO—|—H | H—|—OMe
HO—|—H | HO—|—H
Ph | | Ph

Fischer projections

(b) 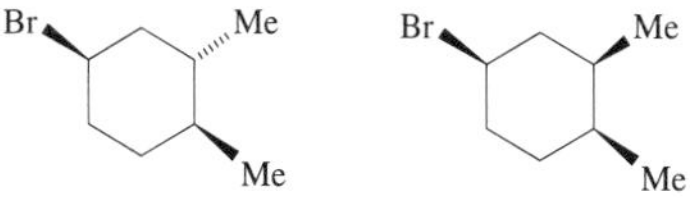 Br, Me, Me; Br, Me, Me

(c) 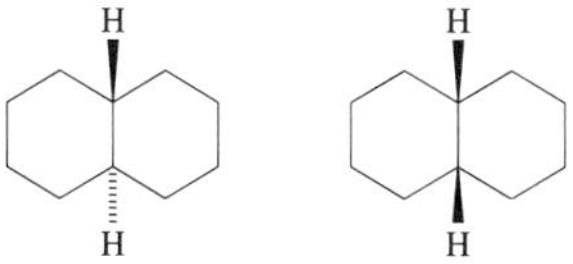 H, H; H, H

25. For **THREE** of the following reactions, which each give a mono-substituted product in a synthetically acceptable yield, write down the structure of the major product and give an explanation for the selectivity observed.

(a) CO_2H, Cl, CO_2H $\xrightarrow{HNO_3/H_2SO_4}$

(b) 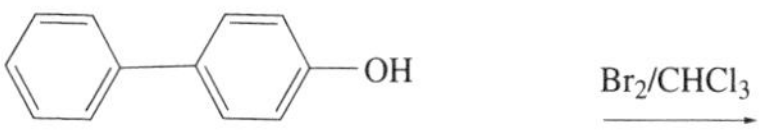 OH $\xrightarrow{Br_2/CHCl_3}$

(c)  $\xrightarrow{HNO_3/CH_3CO_2H}$

(d) 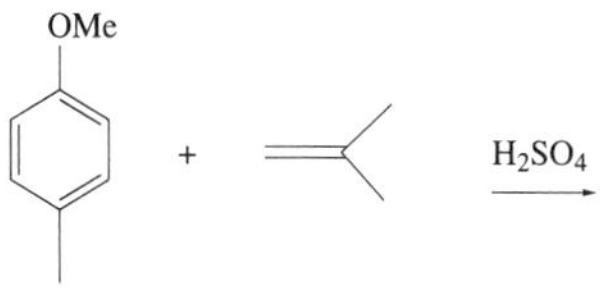 OMe + $\xrightarrow{H_2SO_4}$

(e) 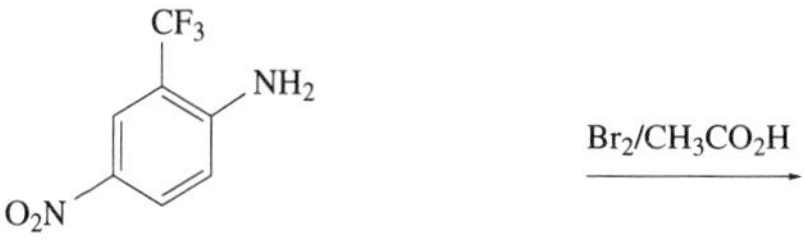 CF_3, NH_2, O_2N $\xrightarrow{Br_2/CH_3CO_2H}$

26. Deduce the structures of the compounds **A** to **G** in the following sequence and comment briefly on the stereochemistry and mechanisms of the reactions.

$C_{10}H_{10}$ (**A**) —[i. $NaNH_2$, liq. NH_3; ii. n-C_4H_9Br]→ $C_{14}H_{18}$ (**B**) —[i. B_2H_6; ii. H_2O_2NaOH]→ $C_{14}H_{20}O$ (**E** and **F** (both))

$C_{14}H_{18}$ (**B**) —[H_2, Pd/$BaSO_4$/S, quinoline]→ $C_{14}H_{20}$ (**C**)

$C_{14}H_{20}$ (**C**) —[i. O_3, MeOH, 0 °C; ii. H_2Pd/C]→ $C_5H_{10}O$ (**D**) + Ph–CH₂–CH₂–CHO

$C_{14}H_{20}$ (**C**) —[Br_2, CCl_4, 0 °C]→ $C_{14}H_2Br_2$ (**G**)

27. Give **ONE** example of the reaction of an electrophilic reagent with (a) 2-butene and (b) benzene. Why do alkenes and benzene differ in their reactions with electrophilic reagents?

28. Write brief notes on the following:

(a) The Diels–Alder reaction between dienes and dienophiles.

(b) The use of ethyl acetoacetate ($CH_3COCH_2COOC_2H_5$) in organic synthesis.

(c) The Michael reaction between carbanions and unsaturated carbonyl compounds, *e.g.* R—CH═CH—CO—R (R = alkyl).

(d) The production of iodoform (CHI_3) when propanone is treated with iodine in aqueous sodium hydroxide solution.

29. (a) Write down the mechanism for a Friedel–Crafts acylation reaction.

(b) Explain the chemistry underlying **TWO** of the following observations.

(i) Nitrobenzene is often used as a solvent in Friedel–Crafts reactions.

(ii) Bromination of aniline (aminobenzene) under mild conditions (bromine in aqueous solution) very rapidly gives a high yield of 2,4,6-tribromaniline, but nitration of aniline with a mixture of conc. nitric and conc. sulfuric acids gives a low yield of 3-nitroaniline in a very slow reaction.

(iii) Bromination of toluene (methylbenzene) gives 2-, 3- and 4-bromotoluene in 55, 2 and 43% yields respectively whilst bromination of *t*-butyl-benzene gives 2-, 3- and 4-bromo-*t*-butyl-benzene in 25, 5 and 70% yields respectively.

30. Define the terms **enantiomer, diastereomer** and **optical activity**. Discuss briefly the stereochemistry of the following compounds:
(a) 2,3-pentanediol.
(b) Tartaric acid (HOOCCH(OH)CH(OH)COOH).

31. Specify reagents suitable for converting 3-ethylpent-2-ene to **EACH** of the following and write brief notes on the mechanisms of **TWO** of these reactions.
(a) 2,3-Dibromo-3-ethylpentane.
(b) 3-Chloro-3-ethylpentane.
(c) 2-Bromo-3-ethylpentane.
(d) 3-Ethyl-3-pentanol.
(e) 2,2-Diethyl-3-methyloxirane.
(f) 3-Ethylpentane.

32. Briefly discuss the methods available for reducing organic compounds.
When compound **A** is treated with a solution of ethanedithiol and a catalytic amount of anhydrous HCl, **B**, $C_{11}H_{14}S_2$, is formed. Reaction of **B**, with Raney nickel in an inert solvent gives C, C_9H_{12}. Given that there is an ethyl group and an aromatic part to the compound, what are the structures of **A, B** and **C**? Explain your reasoning. What is the mechanism for the formation of **B**?

33. Discuss the following statement:
"The alkenes are an important class of compounds in synthetic organic chemistry, as they may be prepared from several other classes of compound and they undergo a wide range of reactions."

34. Define the terms **electrophile, nucleophile** and **functional group interconversion (FGI)**.
Devise syntheses for **TWO** of the following from the starting materials shown.

(a) 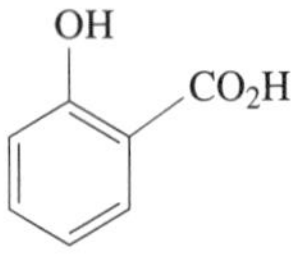 [structure: 2-hydroxybenzoic acid, OH and CO_2H on benzene ring] from [structure: phenol, OH on benzene ring]

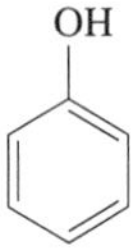

(b) 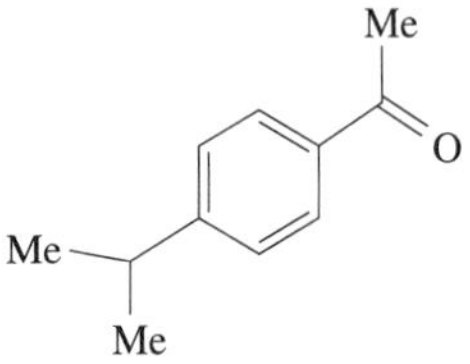 [structure: 4-isopropylacetophenone, Me, O, Me, Me] from [structure: benzene]

(c) $(CH_3)_2CHCH_2COCH_3$ from Me_2CHBr

35. Discuss briefly the Diels–Alder reaction between dienes and dienophiles.

2,3-Dimethylbuta-1,3-diene will react with methyl vinyl ketone ($CH_3COCH{=}CH_2$) to yield compound **A**, $C_{10}H_{16}O$. Treatment of compound **A** with the ylide prepared from triphenyl phosphine and iodoethane (a Wittig reaction) gives compound **B**, $C_{12}H_{20}$.

Deduce structures for compounds **A** and **B** and discuss the reactions involved in their formation.

36. What are the products of the following reactions?

Draw a mechanism for **EACH** transformation and comment on points of stereochemical importance.

(a) 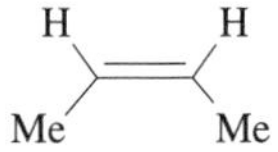 [structure: cis-2-butene, H, H, Me, Me] $\xrightarrow[\text{inert solvent}]{\text{DBr}}$

(b) CH_2CO_2Et—CH_2CO_2Et $\xrightarrow{\text{excess } EtO^- Na^+/EtOD}$

(c) PhCHO $\xrightarrow[\text{in aqueous ethanol}]{HON^+H_3Cl^-/CH_3CO_2^- Na^+}$

37. Devise syntheses for **THREE** of the compounds given below. For **EACH** synthesis the compound indicated should be used as one of the starting materials. More than one step may be needed for each synthesis.

(a) 4-HO–C_6H_4–$NHCOCH_3$ from C_6H_5OH (HO–phenyl)

(b) $(H_3C)_2CHCH_2COOH$ from $(H_3C)_2CHBr$

(c) $CH_3OCH_2CH_2CN$ from $CH_2{=}CHCN$

(d) methylenecyclohexane ($C_6H_{10}{=}CH_2$) from cyclohexanone ($C_6H_{10}{=}O$)

(e) 4,5-dimethylcyclohex-4-ene-1,2-diyl with two CH_2OH groups (H_3C, H_3C, CH_2OH, CH_2OH) from maleic anhydride

38. Using your knowledge of the reactivity of alkenes, alkynes and alcohols describe the reactions that compound **A**, an intermediate in a synthesis of a natural product, might be expected to undergo. Include in your answer mechanisms for **TWO** of the predicted reactions.

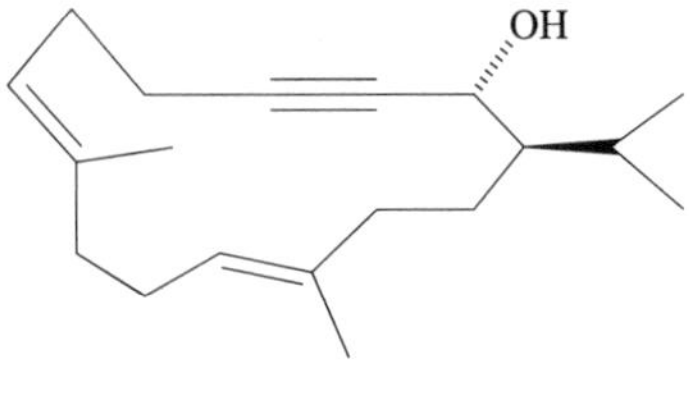

OH

A

39. Compare and contrast the S_N2 and S_N1 mechanisms for nucleophilic substitution.

40. Suggest methods for carrying out **FOUR** of the following conversions.

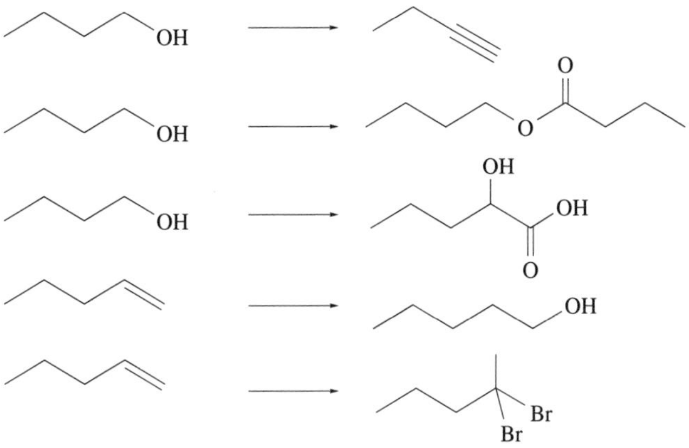

OH ⟶

OH ⟶ O O

OH ⟶ OH OH O

⟶ OH

⟶ Br Br

41. Treatment of compound **A**, C_8H_8O, with zinc/mercury amalgam gave **B**, $C_{16}H_{18}O_2$. Careful dehydration of the latter gave **C**, $C_{16}H_{14}$, which reacted with maleic anhydride **D** to produce **E**, $C_{20}H_{16}O_3$. Treatment of **B** with aqueous acid gave **F**, $C_{16}H_{16}O$, which reacted with iodine and sodium hydroxide to liberate iodoform, CHI_3.

What are the structures of compounds **A**, **B**, **C**, **E** and **F**? Discuss briefly the chemical reactions involved in each of the steps outlined above.

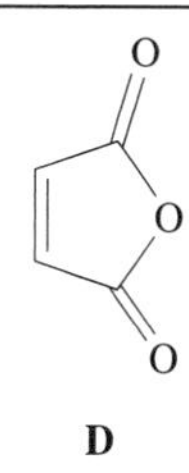

42. Describe the molecular and electronic (orbital) structures of:

(a) Ethane.

(b) The ethanoate (acetate) ion.

(c) Ethyne.

43. (a) Give the absolute configurations of the following compounds in terms of the Cahn–Ingold–Prelog (*R*/*S*) system.

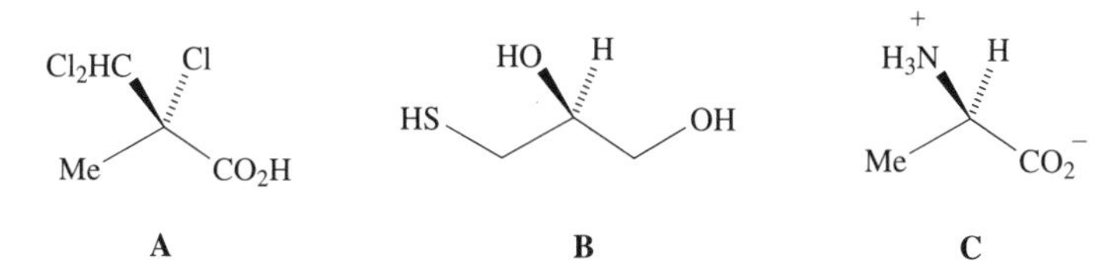

Rewrite the structures **A** to **C** using Fischer projection representation.

(b) (2*R*)-1-Phenylpropan-2-ol, on treatment with PBr_3 and pyridine gave a product **D**, $C_9H_{11}Br$. When **D** was treated with sodium cyanide, a compound **F** was produced, which on hydrolysis with boiling hydrochloric acid gave an acid **F**, $C_{10}H_{12}O_2$.

Give the structures of compounds **D** to **F**, paying particular attention to stereochemistry, and account briefly for the reactions described.

44. Write brief notes on **TWO** methods of preparing alkenes.

Treatment of 2,2,3,4,4-pentamethyl-3-pentanol with concentrated sulfuric acid gave two alkenes **A** and **B**. Ozonolysis of **A** gave formaldehyde and 2,2,4,4-tetramethyl-3-pentanone. Ozonolysis of **B** gave formaldehyde and 3,3,4,4-tetramethyl-2-pentanone.

(a) Identify **A** and **B**.

(b) Suggest a mechanism for the formation of **A**.

(c) Suggest a mechanism for the formation of **B**.

45. Discuss briefly **THREE** of the following:

(a) The electronic structure of aldehydes and ketones.

(b) The mechanism of nucleophilic addition to aldehydes with particular reference to the formation of cyanohydrins (RCHOHCN), acetals ($RCH(OR^1)_2$ and hydrazones ($RCH{=}NNHR^1$).
(c) The oxidation of secondary alcohols to ketones.
(d) Enolisation.

46. Discuss the base-catalysed synthesis of ethyl acetoacetate ($CH_3COCH_2COOC_2H_5$) from two equivalents of ethyl ethanoate. Treatment of benzaldehyde (PhCHO) with ethyl acetoacetate in the presence of sodium ethoxide gives a neutral compound **A**, $C_{13}H_{16}O_4$. Acidic hydrolysis of compound **A** followed by heating produced a ketone **B**, $C_{10}H_{10}O$. What are the structures of compounds **A** and **B**? What reactions might be involved in their formation.

47. Devise syntheses for **THREE** of the compounds given below. For **EACH** synthesis the compound indicated should be used as one of the starting materials.

(a) PhCH=CHCOOH from PhCHO

(b) cyclopropyl–$COCH_3$ from CH_3COCH_2COOEt

(c) H / H_3C C=C H / CH_2Ph (trans-alkene: H_3C–CH=CH–CH_2Ph) from H_3C–C≡C–H

(d) $PhCH(COOEt)_2$ from $PhCH_2COOEt$

(e) 4-CH_3O–C_6H_4–CH_2NH_2 (OCH_3 / CH_2NH_2) from 4-CH_3O–C_6H_4–CHO (OCH_3 / CHO)

48. Explain, with reasoning, which member of the following pairs is the stronger acid (a–b) or the stronger base (c–e) in water:

(a) $O_2NCH_2CO_2H$ $CH_3CH_2CO_2H$

(b) HO_2C CO_2H HO_2C CO_2H

(first dissociations only)

(c) N H N H

(d) NEt_2 O_2N NO_2 NO_2 NH_2 O_2N NO_2 NO_2

(e) N O N O Et

49. Answer parts (i) to (iii).

(i) Give an illustrative example in each case of an organic reaction which involves the following intermediates:

(a) Carbocation.
(b) Carbanion.
(c) Carbon radical.
(d) Carbene.

(ii) Assign the following pK_a values to the following carbon acids:
pK_a values: 43, 37, 33, 25, 20, 13, 4.

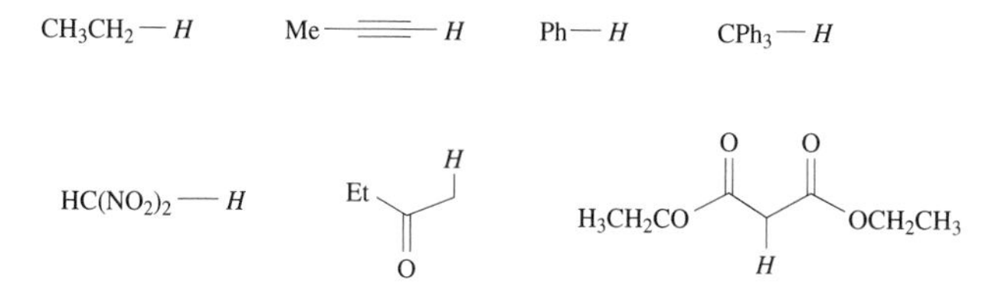

(iii) Assign, with rationalisation, an approximate pK_a value to a different class of carbon acid of your own choice.

50. Answer parts (i) to (iii).

(i) Illustrate the E1 and E2 mechanisms by which hydrogen bromide can be eliminated from alkyl bromides under basic conditions.

(ii) Describe the methods which might be used to establish whether it was the E2 or the E1 mechanism which was predominating in a given elimination reaction.

(iii) Predict, with justification, the structures of the major products, **A** and **B**, of the following reactions:

(a) Cl, $NaOCH_2CH_3$/ $HOCH_2CH_3$/heat → **A** C_7H_{12}

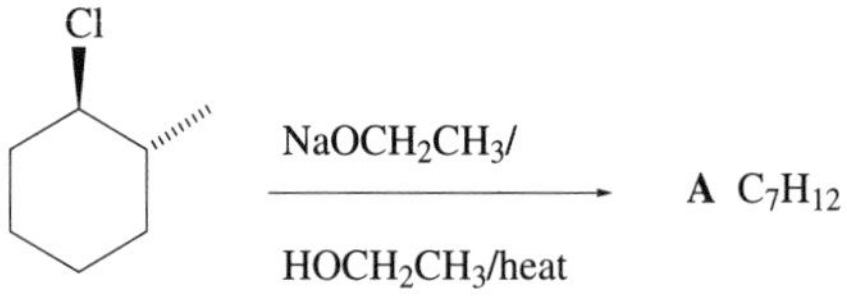

(b) Cl, $NaOCH_2CH_3$/ $HOCH_2CH_3$/heat → **B** C_7H_{12}

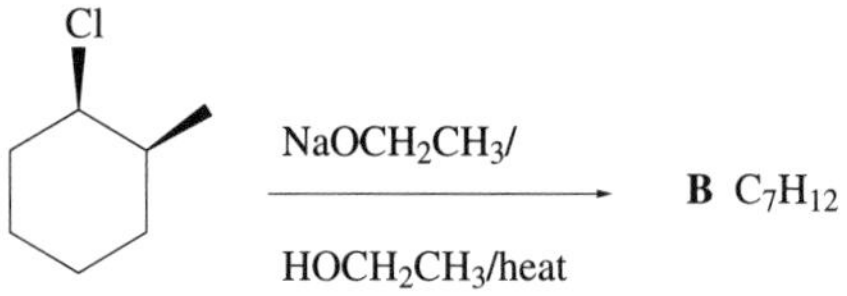

51. Answer parts (i) and (ii).

(i) Draw representations of the isomers (other than conformational) of dichlorocyclopentane.

Use these isomers to illustrate the following terms:

(a) Racemic mixture.
(b) Enantiomer.
(c) Diastereomer.
(d) Achiral.
(e) Meso.

(ii) Identify stereoisomers (other than conformational) of **THREE** of the following compounds:

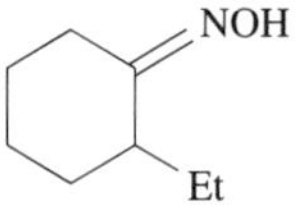

NOH, Et — **A**

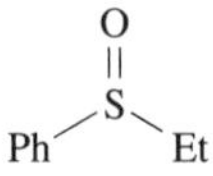

O, S, Ph, Et — **B**

$H_3CHC{=}C{=}CHCH_3$ — **C**

$HClC{=}CBrI$ — **D**

52. Answer parts (i) to (iii).

(i) Discuss the effect of (a) substrate structure and (b) solvent on nucleophilic aliphatic substitution reactions.

(ii) Predict, with reasoning, the faster reaction in each of the following pairs:

(a) Ph_3CCl *versus* 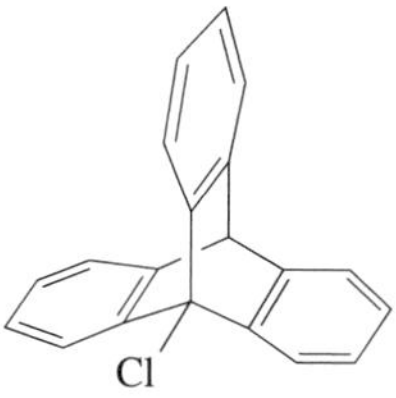 Cl — S_N1 reaction in CH_3CH_2OH at 25 °C

(b) 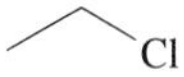 Cl *versus* 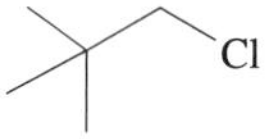 Cl — S_N2 reaction with $NaCH_2CH_3$ in CH_3CH_2OH

(c) 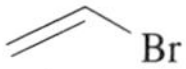 Br *versus*  Br — S_N2 reaction with NaSPh

(iii) Predict, with mechanistic reasoning, the structures of the products of the following reactions:

(a) 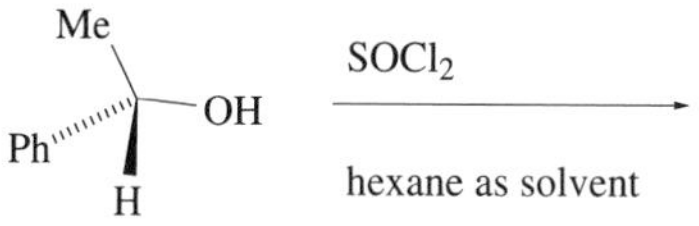 Me, Ph, OH, H $\xrightarrow[\text{hexane as solvent}]{SOCl_2}$ **A** C_8H_9Cl

(b) 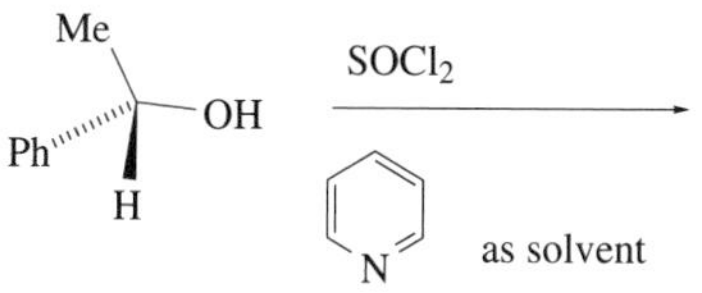 Me, Ph, OH, H $\xrightarrow{SOCl_2}$ **B** C_8H_9Cl

N as solvent

53. Propose synthetic routes and suggest mechanisms for **FIVE** of the following conversions:

(a) Et—≡—H ⟶ (product with Et, OH)

(b) HO(C=O)(CH₂)₄(C=O)OH ⟶ 2-ethylcyclopentanone (O, Et)

(c) 4-methylcyclohexanone (O) ⟶ methylidene-4-methylcyclohexane

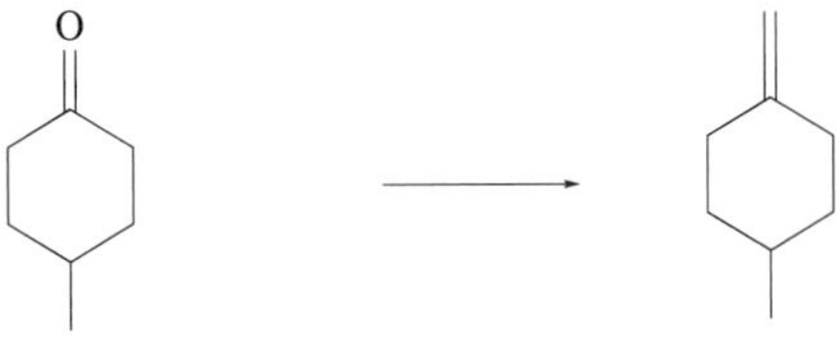

(d) Et—≡—Et ⟶ Et, OH / Et, OH (racemic)

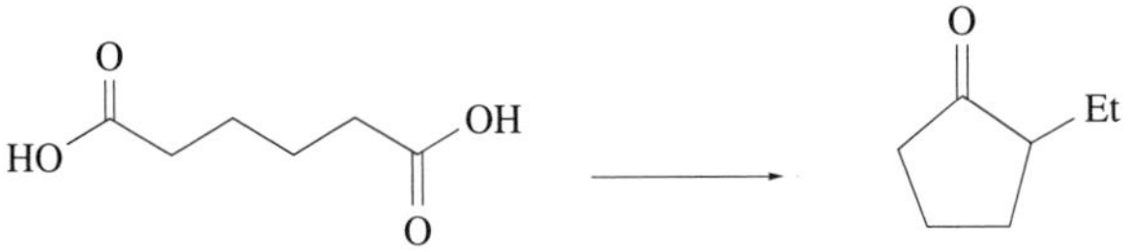

(e) Et₃C—OH ⟶ Et₃C—CO_2H

(f) OH ⟶ NH_2 (racemic)

(g) Et, Et ⟶ Et, H, CCl_2, Et, H

54. Answer parts (i) to (iv).

(i) Explain the following order of reaction rates with bromine in the dark:

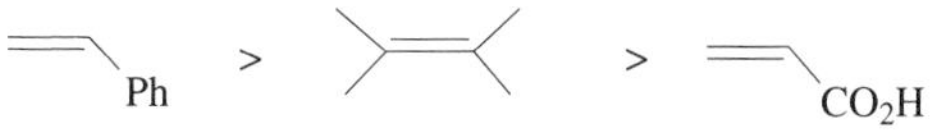

Ph > > CO_2H

(ii) Explain the following rate order for nucleophilic substitution reaction with aqueous ammonia:

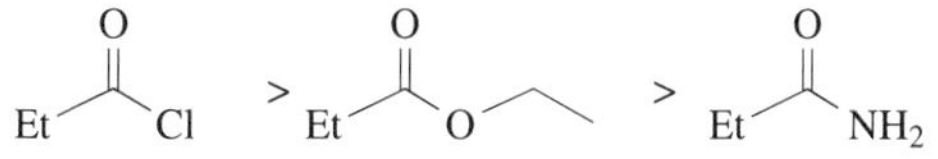

O O O
Et Cl > Et O > Et NH_2

(iii) Illustrate, with an example of your own choice, the term "kinetic versus thermodynamic control" as applied to an organic reaction.

(iv) Explain the following observations:

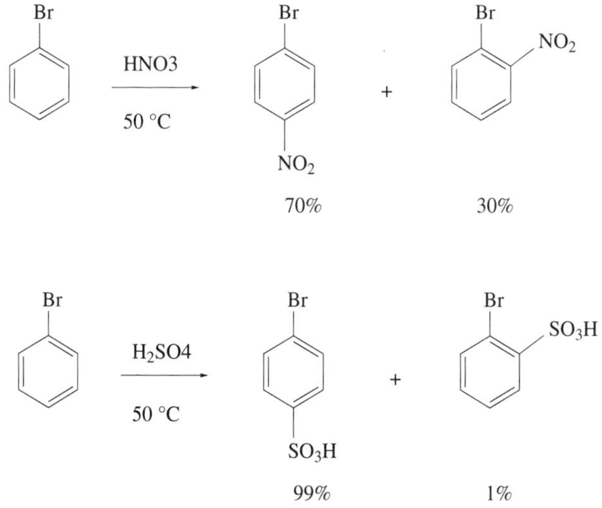

Br — HNO3, 50 °C → Br, NO_2 (70%) + Br, NO_2 (30%)

Br — H_2SO4, 50 °C → Br, SO_3H (99%) + Br, SO_3H (1%)

55. Answer parts (i) and (ii).

(i) Provide mechanistic rationalisation for **FIVE** of the following conversions:

(a) 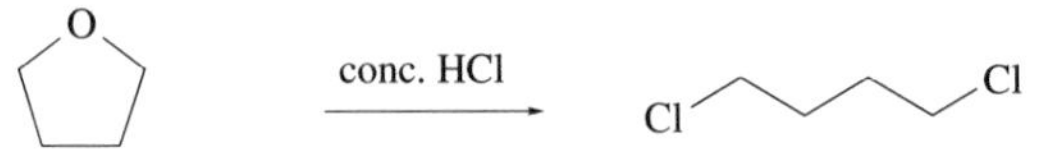 O (tetrahydrofuran) $\xrightarrow{\text{conc. HCl}}$ Cl–(CH₂)₄–Cl

(b) 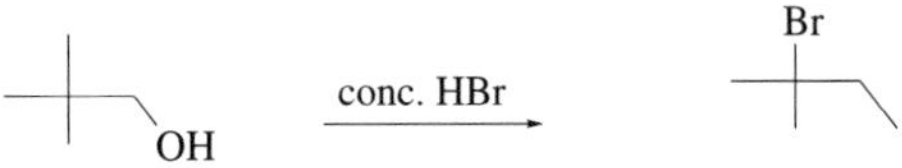 OH $\xrightarrow{\text{conc. HBr}}$ Br

(c) 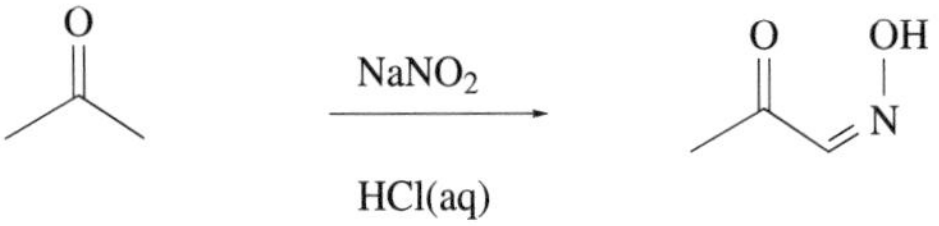 O $\xrightarrow[\text{HCl(aq)}]{\text{NaNO}_2}$ O OH N

(d) 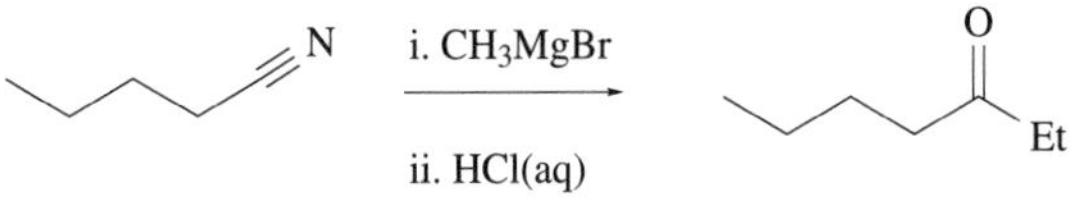 N $\xrightarrow[\text{ii. HCl(aq)}]{\text{i. CH}_3\text{MgBr}}$ O Et

(e) 2 PhCHO $\xrightarrow[\text{ii. HCl(aq)}]{\text{i. conc. HO}^-}$ $PhCH_2OH$ + $PhCO_2H$

(f) 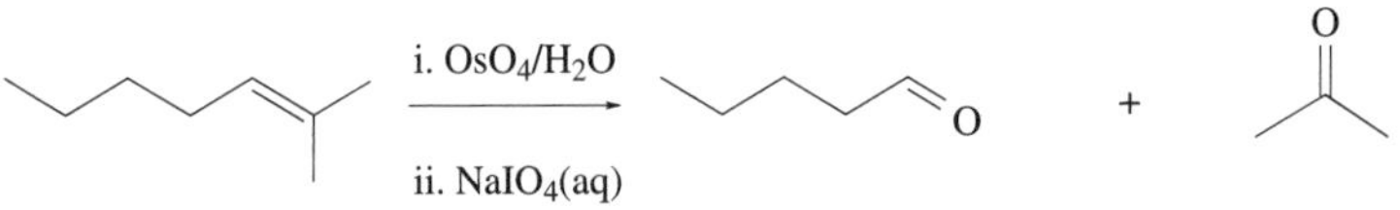 $\xrightarrow[\text{ii. NaIO}_4\text{(aq)}]{\text{i. OsO}_4\text{/H}_2\text{O}}$ O + O

(g) 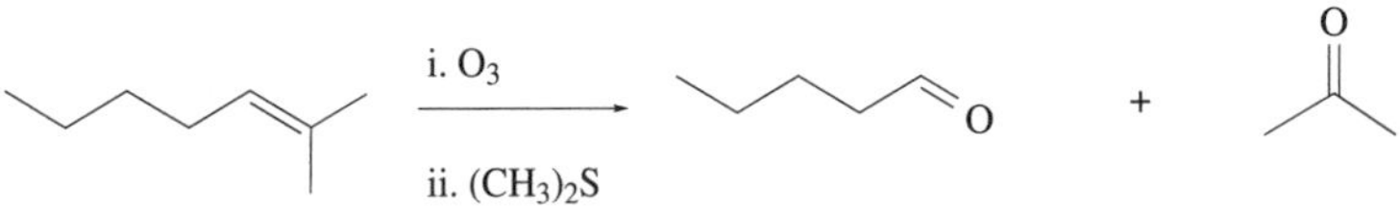 $\xrightarrow[\text{ii. (CH}_3)_2\text{S}]{\text{i. O}_3}$ O + O

(ii) Propose a synthesis of **ONE** of the following isotopically labelled compounds using $^{13}CO_2$ as the source of label:

(a) 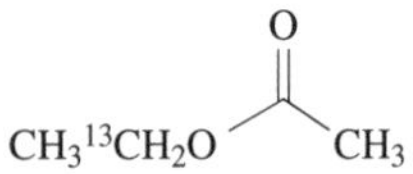 $CH_3{}^{13}CH_2O$–C(=O)–CH_3

(b) 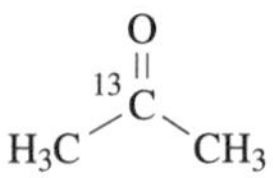 H_3C–^{13}C(=O)–CH_3

56. Design a route for **FOUR** of the following syntheses, which all require more than one step to form the product.

(a)  MeO_2C–CH_2–CO_2Me ⟶ HO_2C–CH_2–CH_2Ph

(b)  cyclopentyl–Br ⟶ spiro epoxide (O, Ph)

(c) 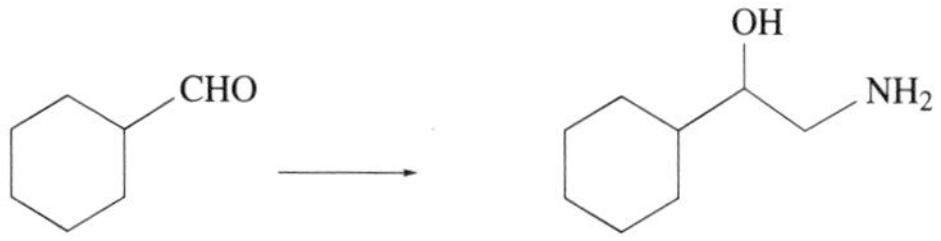 cyclohexyl–CHO ⟶ cyclohexyl–CH(OH)–CH_2–NH_2

(d)  C_7H_{15}–C≡CH ⟶ C_7H_{15}–cyclopropyl

(e)  methylenecyclohexane ⟶ cyclohexyl–CH_2–OMe

(f)  C_6H_{14}–C(=O)–NH_2 ⟶ C_6H_{14}–CH_2–NH_2

57. (a) Give examples of the formation of enolates from ketones and esters. What factors determine whether the kinetic or thermodynamic enolate is formed preferentially?
(b) Give examples of compounds which react with enolates, highlighting any mechanistic or synthetic points of interest.

58. Design syntheses of the following molecules, using starting materials which contain no more than four carbon atoms.

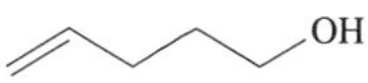

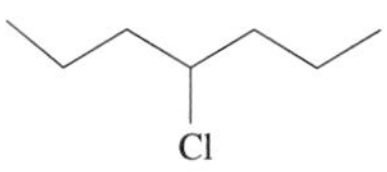

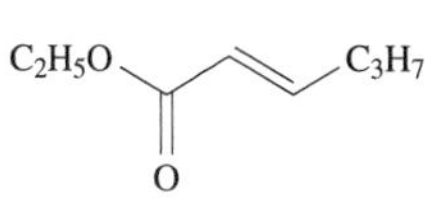

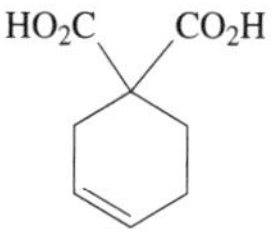

OH; Cl; C_2H_5O, O, C_3H_7; HO_2C, CO_2H

59. (a) Using cyclohexanone as an example, briefly discuss the reactions of ketones with:

(i) Hydroxylamine.

(ii) Secondary amines (R_2NH).

(iii) Zinc and hydrochloric acid.

(iv) Trimethylsilyl cyanide (Me_3SiCN).

(b) Compare the rate of nitration (nitric and sulfuric acid) of toluene, thiophene and pyridine. Discuss the mechanism, the stability and structure of the intermediates, and the regioselectivity of the nitration of toluene and thiophene.

60. Starting from benzene or toluene, indicate how the following compounds can be synthesised:

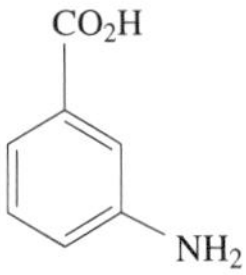

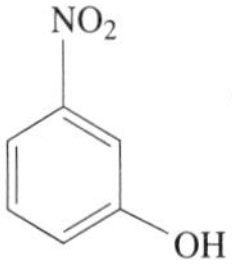

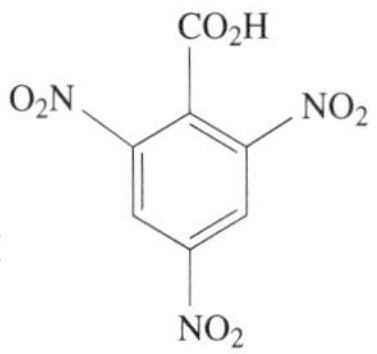

61. Discuss the use of aromatic diazonium salts in the synthesis of substituted benzenes.

62. Describe the chemistry of:

(a) Aliphatic dicarboxylic acids.

(b) Aliphatic β-keto acids and their derivatives.

63. Draw the following structures in their most stable conformations. In each case say whether the molecule represented would be optically active, and whether it would have diastereomers (indicating briefly how their structures would differ from the original).

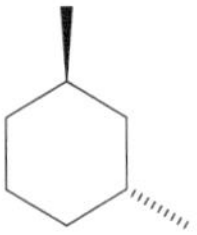

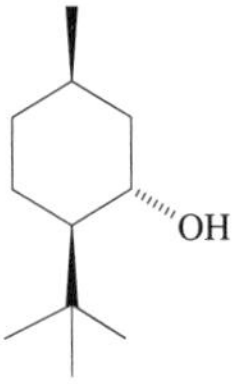

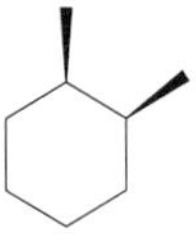

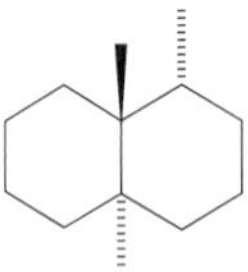

$CH_3CH_2CH_2CH(CH_3)_2$

64. Distinguish between the E1 and E2 mechanisms giving a typical example of a reaction which demonstrates each mechanism.

Complete the following reactions showing all possible alkene products.

(a) 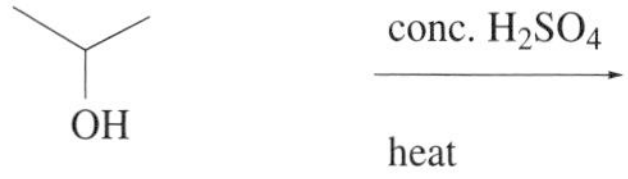 OH $\xrightarrow[\text{heat}]{\text{conc. } H_2SO_4}$

(b) 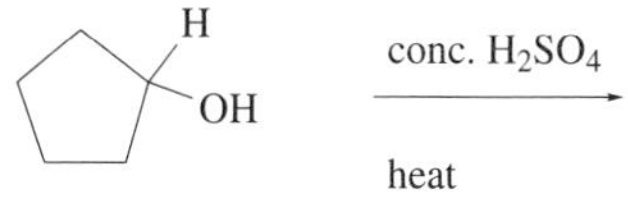 H OH $\xrightarrow[\text{heat}]{\text{conc. } H_2SO_4}$

(c)  I $\xrightarrow{\text{heat}}$

65. A dipeptide below can be obtained by partial hydrolysis of proteins.

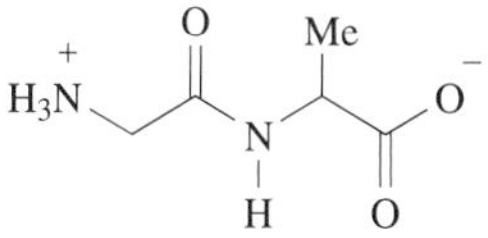

$\overset{+}{H_3N}$ O Me $\overset{-}{O}$ N H O

Give the structures of the products that would be obtained when the dipeptide is treated with each of the following sets of reagents:

(a) NaOH, H_2O, heat.
(b) H_2SO_4, H_2O, heat.
(c) (i) $LiAlH_4$; (ii) H_2O, H^+.

66. Predict the products from the following reactions and give the mechanism for any one example.

(a) $(CH_3)_2CuI + \overset{+}{K}\ \overset{-}{CN} \longrightarrow$

(b) $CH_3I + CH_3CH_2\overset{-}{S}\ \overset{+}{Na} \longrightarrow$

(c) 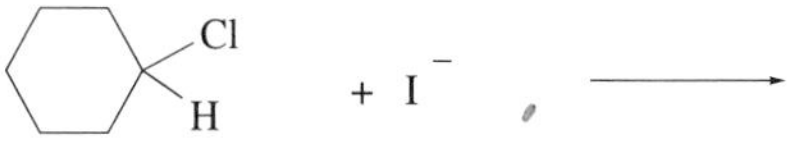 Cl H $+ \ \overset{-}{I} \longrightarrow$

67. Explain what is meant by the following terms and give an example in support of your answer:
(a) Halogenation.
(b) Hydrogenation.
(c) Hydration.
(d) Hydrolysis.

68. Discuss **THREE** of the following reactions from the point of view of the mechanism and the stereochemistry of the product(s).

(a) cyclohexene — i. RCO_3H; ii. $\overset{+}{H}/H_2O$ →

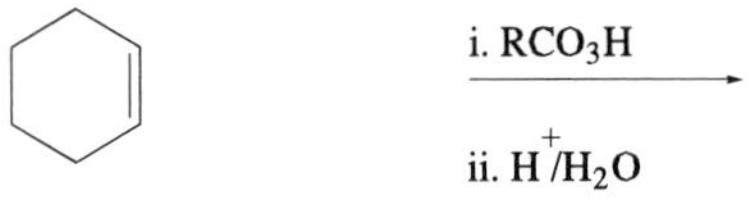

(b) R—≡—R — Na in liquid ammonia →

(c) H_3C, C_2H_5, C_3H_7, Br — $CH_3CO_2^-$ in C_2H_5OH →

(d) O, OMe, OMe, O →

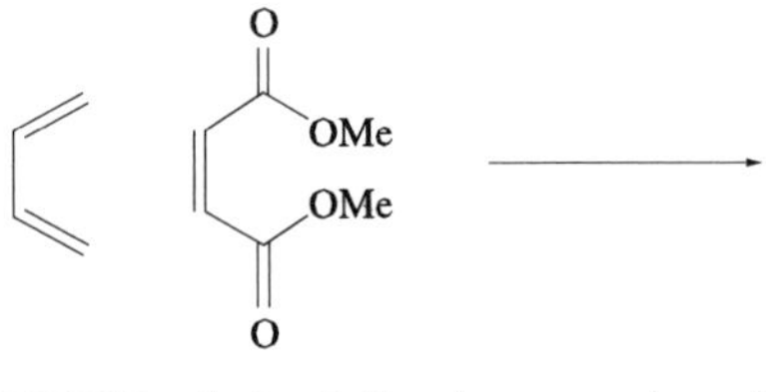

69. Explain the use of **FOUR** of the following reactions in synthetic organic chemistry, and discuss them from a mechanistic point of view.
(a) Alkenes with B_2H_6 followed by H_2O_2/NaOH.
(b) Primary alcohols with $POCl_3$ and pyridine.
(c) Ketones with the ylide tautomer of $PhP{=}CH_2$.
(d) Haloalkanes with BuLi and CuI.
(e) Grignard reagents with an epoxide.
(f) Aldehydes with base.

70. Answer **ALL** parts.
(a) What stereochemistry (*E* or *Z*) would you expect for the alkene obtained by E2 elimination of (1*R*,2*R*)-1,2-dibromo-1,2-diphenylethane? Draw a Newman projection of the reacting conformation.

(b) What products would you expect to obtain from S_N1 reaction of (*S*)-3-chloro-3-methyloctane with sodium acetate in acetic acid? Show the stereochemistry of both starting material and products.

(c) Describe the difference between transition state and reaction intermediate.

(d) The carbonyl groups in esters and amides are less reactive than those in ketones. Explain this difference with the aid of resonance structures.

71. Answer **ALL** parts.

(a) Explain, with the aid of a drawing, how hydrogenation of alkynes produces (*Z*)-alkenes.

Me OH Me

(b) Draw the preferred conformation of the molecule shown below:

$$(Me)_3COMe \xrightarrow{\text{cold conc. HCl}} (Me)_3CCl + MeOH$$

HO Me Br H Me H $\xrightarrow[\text{EtOH}]{\text{NaOEt}}$ O H H Me Me

(c) Carbocations often rearrange by 1,2-shifts. Illustrate this statement with an example where treatment of an alcohol with concentrated HBr leads to a rearranged bromoalkane. What is the major driving force in these rearrangements?

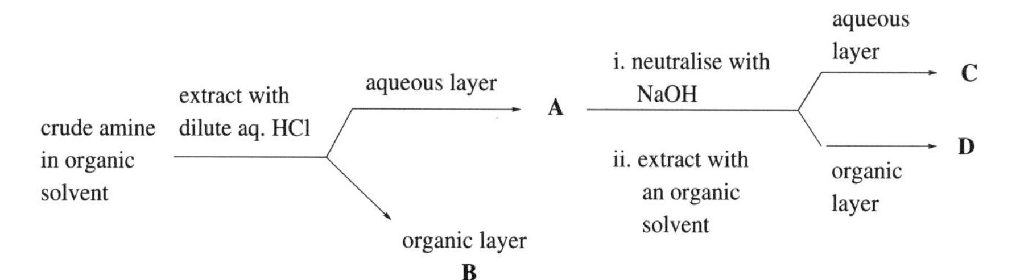

(d) Write mechanisms for the following reactions:

(e) The amide PhCONHMe was reduced by $LiAlH_4$. The crude product in Et_2O, which might have contained some starting material and inorganic salts, was worked up by the procedure outlined in the flow chart below. Explain how pure amine is isolated by this procedure and suggest compositions of **A, B, C** and **D**.

72. Answer **BOTH** parts.

(a) A compound **A** (C_7H_8O), which was insoluble in aqueous base, reacted with Br_2 and $FeBr_3$ to give predominantly **B** (C_7H_7BrO). Treatment of **B** with magnesium (in *dry* diethyl ether), followed by addition of CO_2 gave, after workup, **C** ($C_8H_8O_3$) which was soluble in aqueous base. Reaction of **C** with ethanol in the presence of an acid catalyst gave **D** ($C_{10}H_{12}O_3$) which reacted with $LiAlH_4$ to give, after workup, 4-methoxybenzyl alcohol **E**.

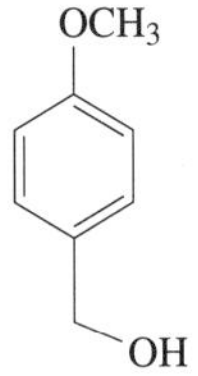

OCH_3

E

OH

Identify the compounds **A** to **D** and give a mechanism for the formation of **D**.

(b) Show all reagents required for **EACH** of the following conversions. Note that more than one step may be needed.

(i)

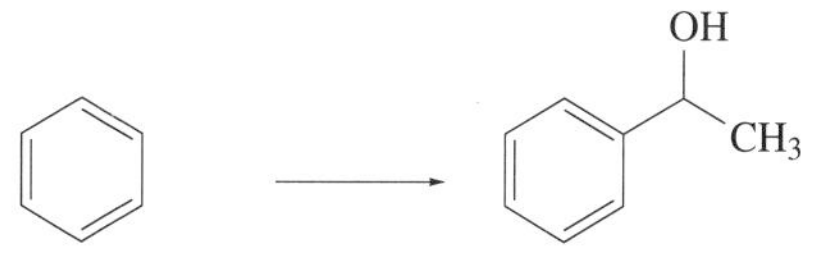

(ii)

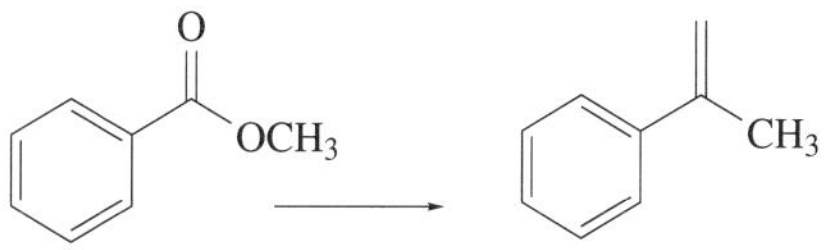

(iii)

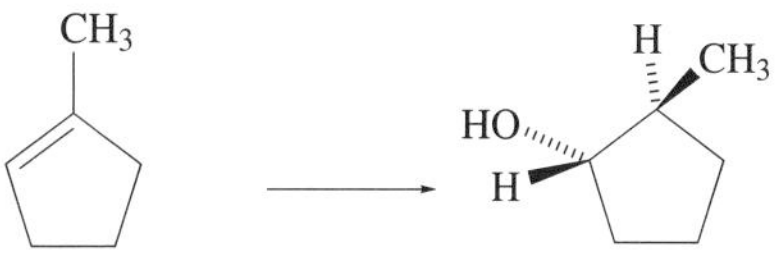

73. For each of the following, give **ONE** example and discuss from a mechanistic point of view:

(a) A concerted reaction.

(b) A reaction involving homolytic steps.

(c) A reaction between aldehydes and a nitrogen nucleophile.

74. Discuss **TWO** of the following:

(a) Carbon–carbon bond formation using enolate anions from aldehydes and ketones.

(b) Reactions of carboxylic acid derivatives with nucleophiles.

(c) Conformation of linear and cyclic alkanes.

75. For **THREE** of the following sequences, give suitable reagents for each step, and discuss the mechanism of the step marked *:

(a)

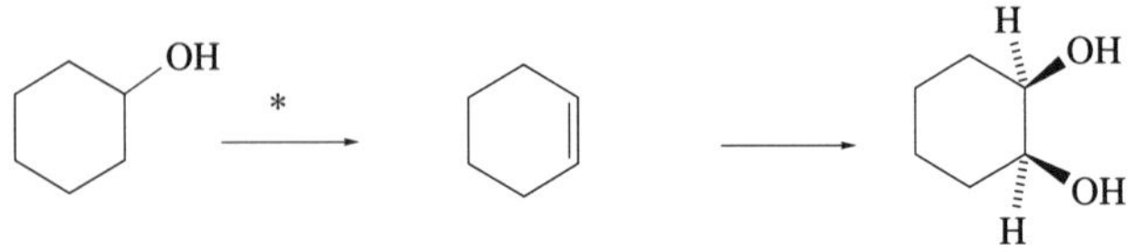

(b)

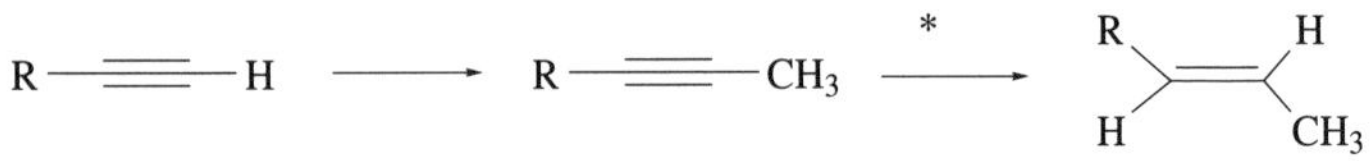

(c)

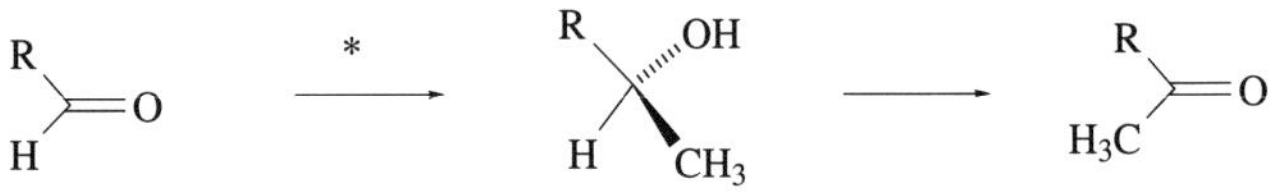

(d)

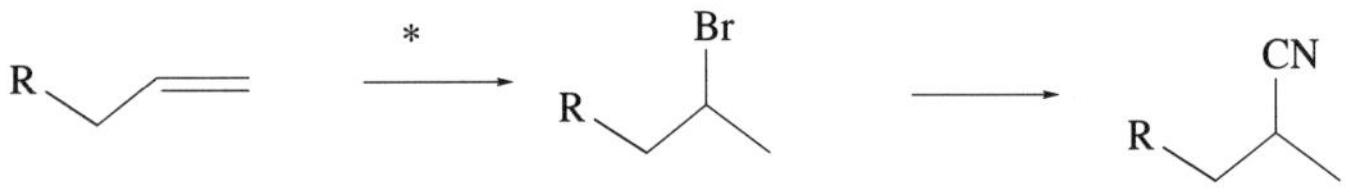

76. (a) Explain in detail why aldehydes are more reactive than ketones towards nucleophiles.
(b) Name the following aldehyde and ketone using IUPAC nomenclature.

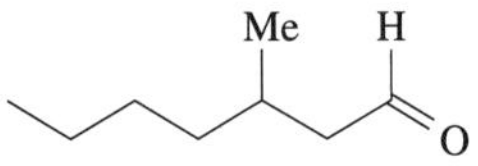

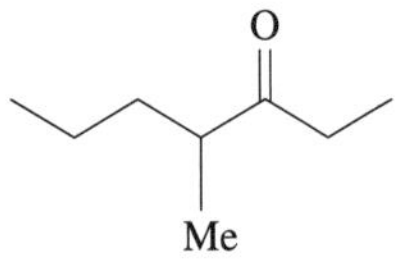

(c) How would you chemically distinguish between the following compounds?
(i) An aldehyde and an ester.
(ii) A ketone and an alkene.
(iii) A carboxylic acid and an ester.

77. (a) Describe in detail how you would separate a mixture of liquid alkane and a carboxylic acid into its pure components.
(b) Why are acyl chlorides more reactive than carboxylic acids towards nucleophiles. Predict the products of the following reactions.

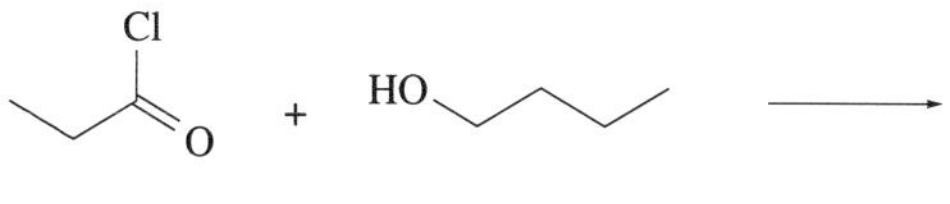

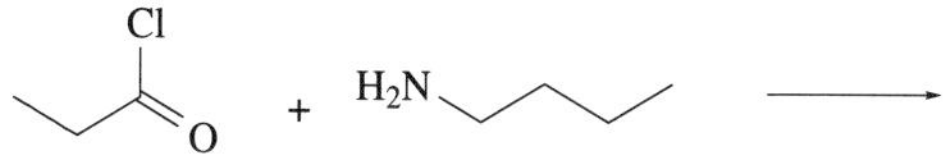

Cl, O, +, HO, →

Cl, O, +, H_2N, →

78. Explain why nitrobenzene is around one million times less reactive than benzene towards electrophilic substitution and undergoes substitution in the 3-position and why chlorobenzene is around thirty times less reactive than benzene towards electrophilic substitution and undergoes substitution in the 2- and 4-positions.

Outline the reagents and essential conditions which you would use to carry out the following sequence of reactions:

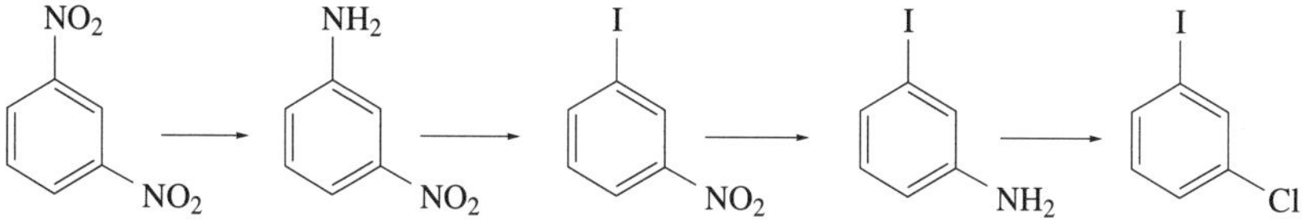

NO_2, NO_2 → NH_2, NO_2 → I, NO_2 → I, NH_2 → I, Cl

79. Attempt **TWO** parts.

(a) Name the following six compounds using the IUPAC system.

$CH_3(CH_2)_2C(CH_3)_3$

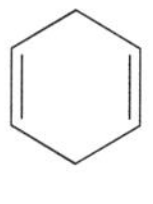

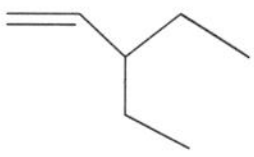

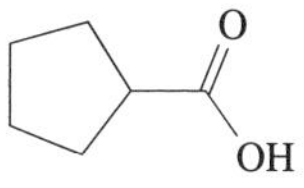

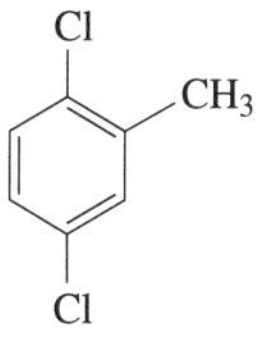

O, OH

Cl, CH_3, Cl

OH

(b) Give the structural formula for each of the following six compounds:

(i) Nonane.
(ii) 2-Methyl-1-butene.
(iii) 1,3-Cyclohexadiene.
(iv) 2-Methyl-2-propanol.
(v) 2,4,6-Tribromophenol.
(vi) D-Glucose.

(c) Give brief answers to each of the following six questions:
  (i) Give the atomic number and the atomic mass of three isotopes of carbon.
  (ii) Use their electronegativity values to rank the elements C, Cl, B, O, and H in ascending order.
  (iii) Describe a covalent bond.
  (iv) What is sp^3 hybridisation?
  (v) Outline the mechanism of *ortho* substitution in benzene.
  (vi) What is a racemic mixture?

80. (a) Describe with diagrams the electrophilic addition of HI to 1-methylcyclopentene. Give the IUPAC name for the product. What rule governs the outcome of this reaction?
(b) Give the structural formula of the monomer, a general formula of the polymer and a general reaction mechanism for the formation of polyvinylchloride.
(c) Describe with diagrams the role of the hydroxyl group in bromination of phenol by electrophylic aromatic substitution.

81. (a) Write a short essay on hydrogen bonding in alcohols. Outline the process for the industrial production of alcohol.
(b) Describe the preparation of a Grignard reagent and how it can be used to synthesise a secondary alcohol.
(c) Give an outline of nuclear magnetic resonance spectroscopy (NMR).

82. Indicate why the following routes are not satisfactory for the synthesis of products **A** and **B**:

CH_3 HNO_3, H_2SO_4 CH_3 NO_2 [O] COOH NO_2 **A**

NO_2 CH_3COCl $AlCl_3$ NO_2 $COCH_3$ **B**

Suggest alternative methods of synthesis for **A** and **B**.

83. "The aromatic ring confers unusual properties on substituents attached to it."
Discuss with reference to OH and NH_2 groups as substituents.

84. Write an essay on the use of diethyl malonate and ethyl acetoacetate (3-oxobutanoate) in synthesis.

85. (a) For formulae (i) to (iii) below, state whether the molecules depicted would be *R* or *S* and give an indication of your reasoning.

(i) 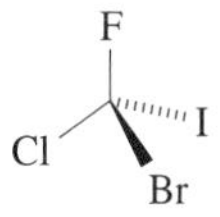 F, Cl, I, Br

(ii) 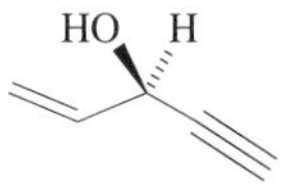 HO H

(iii) 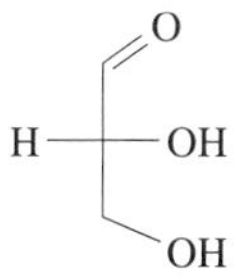 H —— OH, O, OH

(Fischer projection)

(b) For formulae (iv) to (v) below, state whether the molecules depicted would have enantiomers. Where this is the case you should draw the enantiomer, and where it is not you should account for the fact (referring to symmetry elements where relevant).

(iv) 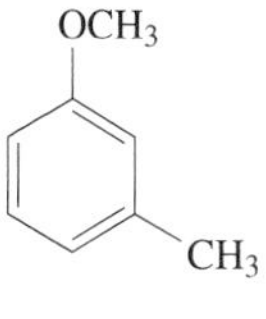 OCH_3, CH_3

(v) 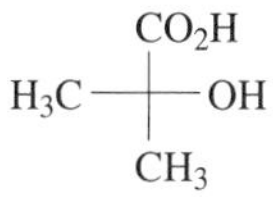 CO_2H, H_3C —— OH, CH_3

86. (a) Name the following compounds.

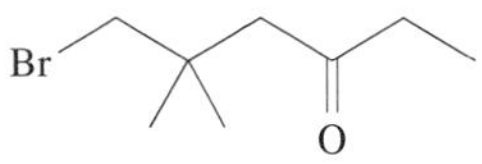

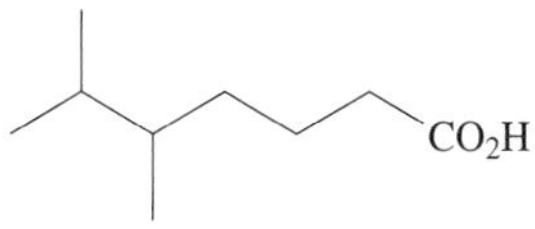

Br, O

CO_2H

(b) Suggest an organic product and indicate a mechanism for each of the following reactions.

(i) [cyclohexanone] + NH_2OH $\xrightarrow{H^+}$

(ii) [cyclohexanone] + MeMgI $\xrightarrow[\text{by } H^+]{\text{followed}}$

(iii) MeO–C(=O)–CH_2CH_3 + NaOH ⟶

87. (a) What is meant by the term "enol–keto tautomerism". Hence explain why ketones react with bromine and suggest the structure of the product obtained.

(b) How could the following alcohols be made from aldehydes, ketones and any other materials.

OH OH OH

Comment on the mechanisms of the reactions employed.

88. Answer **ALL** parts.

Rationalise the following observations:

(a) In their NMR spectra, the methyl group in propanone (acetone) resonates at about $\delta 2.0$ whereas the methyl group in methanol resonates at about $\delta 3.3$.

(b) The carbonyl C=O stretching frequency in the IR spectrum of cyclohexanone is about $1720\,\text{cm}^{-1}$, whilst that in cyclobutanone is about $1780\,\text{cm}^{-1}$.

(c) Two isomers of 1,2-dichloroethene are known. One has a dipole moment of 2.95 D, whereas the other has a dipole moment of zero.

(d) 4-Nitrophenol is much more acidic than 4-methoxyphenol.

(e) In the following compound, one nitrogen is more basic than the other two nitrogens. (Hint: Draw out the various resonance structures to determine which N is the most basic.)

H, Me–N–C(=NH)–NH_2

89. Answer **BOTH** parts.

(a) Very briefly discuss the distinguishing features of substitution reactions which proceed by S_N1 or S_N2 reactions, highlighting the key differences between the two mechanisms.

Use your knowledge to predict the result of the following substitution reactions:

Br, Me — NaCN →

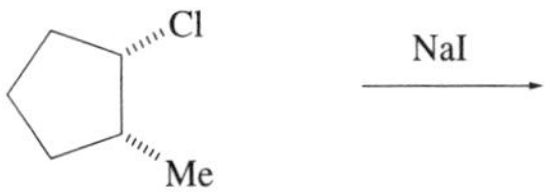

Cl, Me — NaI →

I — $NaNO_2$ →

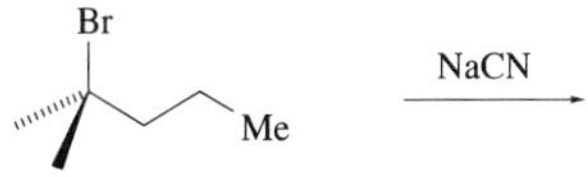

(b) Very briefly discuss the distinguishing features of E2 elimination reactions.

Explain why *cis*-4-*tert*-butylcyclohexyl bromide undergoes elimination (when treated with base) 500 times faster than the *trans*-isomer. What is the product in each case. (Hint: You must draw out the cyclohexane ring in its correct conformation.)

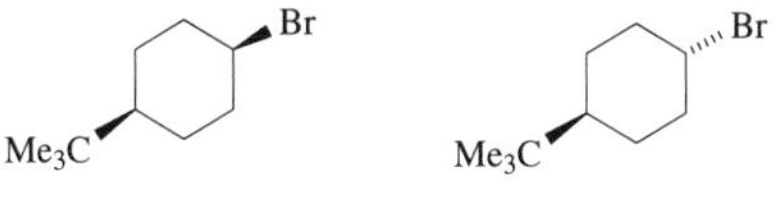

Me_3C, Br — *cis* isomer

Me_3C, Br — *trans* isomer

90. Complete the following reactions:

(a) Cl_2 in MeOH →

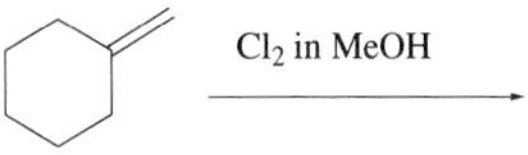

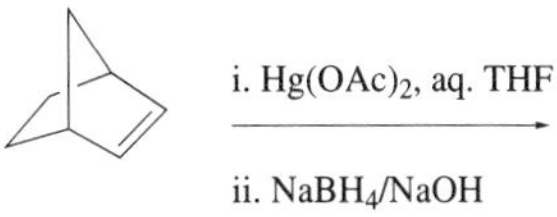

(b) Me — i. 9-BBN/THF; ii. H_2O_2/NaOH →

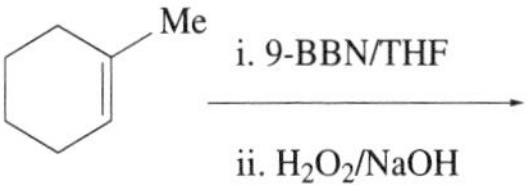

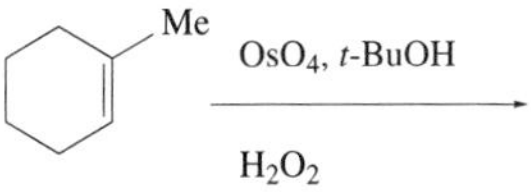

(c) Me — i. O_3; ii. Me_2S →

(d) i. $Hg(OAc)_2$, aq. THF; ii. $NaBH_4$/NaOH →

(e) Me — OsO_4, *t*-BuOH; H_2O_2 →

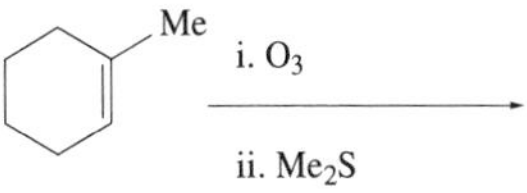

(f) + O, O, O — 0 °C →

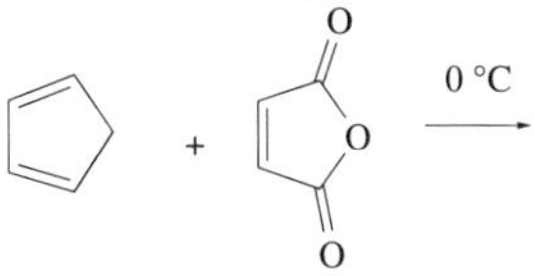

Comment briefly on important mechanistic and stereochemical aspects of reactions (c), (d) and (f).

91. (a) Explain fully the directing effects of nitro and chloro groups in:
(i) Aromatic electrophilic substitution.
(ii) Aromatic nucleophilic substitution.
(b) Predict the monosubstitution products and fully explain the mechanisms for each of the following reactions of *p*-chloronitrobenzene.

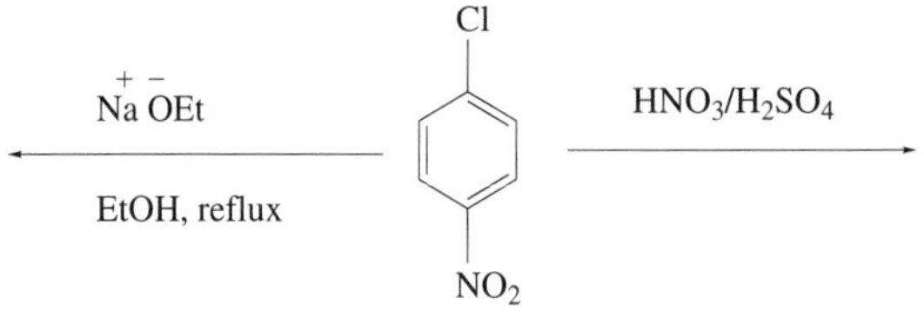

$\overset{+}{Na}\,\overset{-}{O}Et$, EtOH, reflux ← p-chloronitrobenzene (Cl, NO_2) → HNO_3/H_2SO_4

92. (a) With reference to benzene explain what you understand by "resonance" in chemistry.
(b) Show detailed resonance structures for ozone (O_3) and azide ion (N_3^-).

93. (a) Describe, with the aid of examples, (i) the E1 and (ii) the E2 mechanisms for elimination reactions.
(b) When 3-bromo-2,2-dimethylbutane A is heated with a dilute solution of CH_3CH_2ONa in CH_3CH_2OH the reaction follows first-order kinetics and the products **B, C** and **D** are formed. Explain the formation of all the products.

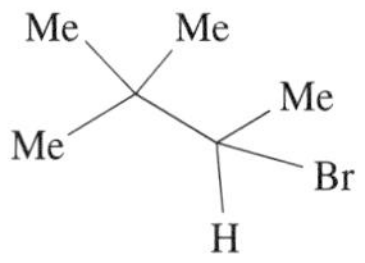

A: Me, Me, Me, Me, Br, H

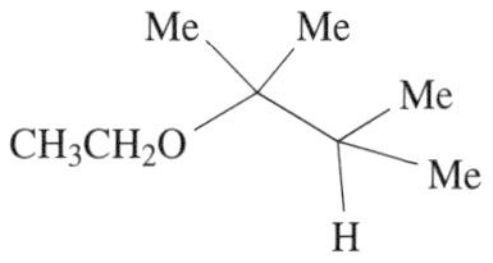

B: Me, Me, CH_3CH_2O, Me, Me, H

C: Me, Me, Me, Me

D: H, Me, Me, Me

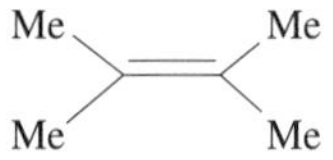

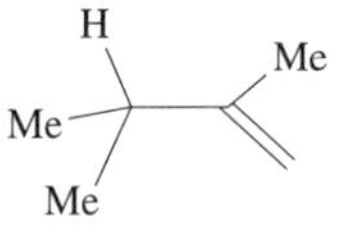

94. (a) Describe the mechanisms of the addition reactions of but-1-ene ($CH_3CH_2CH{=}CH_2$) with (i) bromine and (ii) hydrogen bromide. In the reaction with hydrogen bromide what would be the effect of adding some peroxide to the reaction mixture?

(b) As shown in the following scheme the reaction of 3-methylbut-1-ene (**1**) with hydrogen chloride gives a mixture of two isomeric compounds, (**2**) *and* (**3**). When treated separately with lithium aluminium hydride, (**2**) and (**3**) each give the same alkane (**5**) (2-methylbutane). However, when treated separately with lithium aluminium deutride, (**2**) gives (**4**) and (**3**) gives (**6**). Suggest structures for (**2**) and (**3**) and explain how they formed from (**1**).

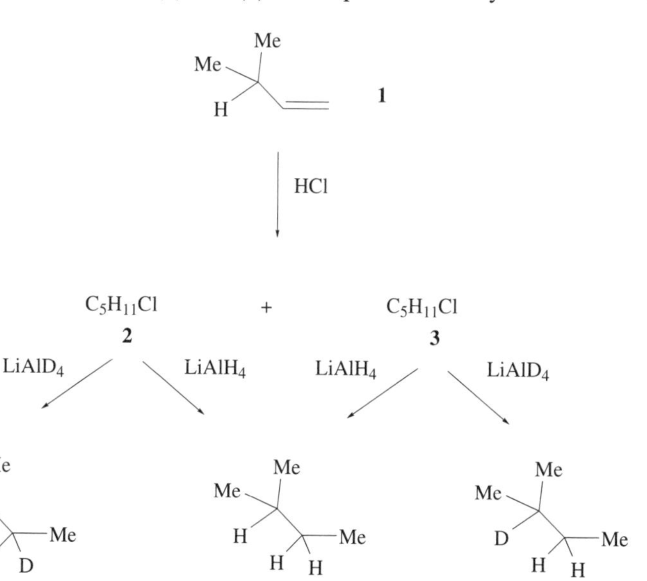

95. (a) Give reagents and conditions required to carry out each of the following conversions and **FOR REACTION (ii) ONLY**, include a description of the mechanism of the reaction that you suggest.

(i)  OH ⟶ O, OH

(ii) 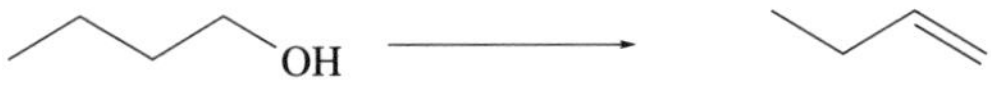 OH ⟶

(iii)  OH ⟶ Br

(iv)  OH ⟶ O

(v)  Me—OH ⟶ O, Me, O

(b) Explain clearly, with appropriate examples, what you understand by the following:

(i) Sequence rules.
(ii) Optical activity.
(iii) Racemic mixture.
(iv) Chiral centre.

96. (a) Suggest reagents for each of the following transformations:

(i)  cyclohexanone ⟶ 2-methylcyclohexanone (O, Me)

(ii)  cyclohexanone ⟶ 1-ethylcyclohexanol (OH, Et)

(iii)  acetone (O) ⟶ HO, CN

(iv) $CH_3CH_2CO_2H \longrightarrow CH_3CH_2COCH_3$

(v) 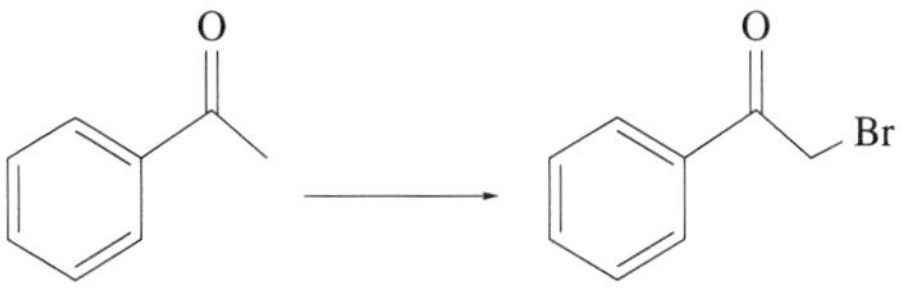 acetophenone (O) ⟶ (O, Br)

(b) Write mechanisms for each of the following reactions using balanced chemical equations and the curved arrow symbolism.

(i) $CH_2(CO_2Et)_2$ + NaOEt + $CH_3Br \longrightarrow CH_3CH(CO_2Et)_2$

(ii) 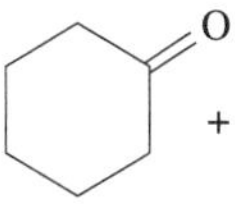 cyclohexanone (O) + 2 CH_3OH + H^+ (catalyst) ⟶ (OCH_3, OCH_3)

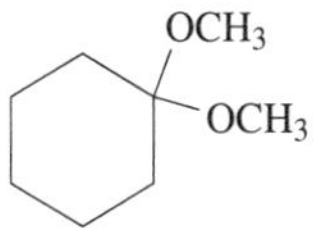

97. (a) Arrange the following compounds in order of decreasing pK_a values (increasing acidity).

CH_3CO_2H CH_3NH_2 CH_3COCH_3 CH_4 CH_3CH_2OH CF_3CO_2H

(b) What is (are) the principal organic product(s) of each of the following reactions:

(i) [bromobenzene] + HNO_3 + H_2SO_4 $\longrightarrow$

(ii) PhCHO + NH_2OH $\longrightarrow$

(iii) CH_3CH_2CHO + $NaBH_4$ $\longrightarrow$

(iv) $CH_3CO_2CH_3$ + CH_3MgBr $\longrightarrow$

(v) $CH_3CH_2CO_2CH_3$ + NaOH followed by H_3O^+ $\longrightarrow$

98. (a) Using the Hückel rule of aromaticity predict which of the following molecules can be classified as aromatic. Briefly give your reasons.

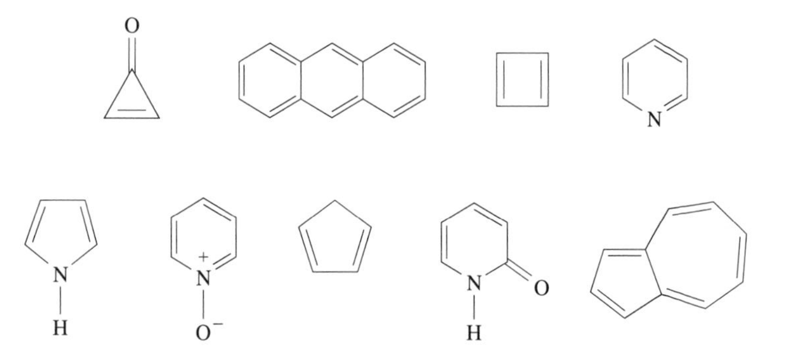

(b) Label the following molecules as *R*, *S*, *E* or *Z*, as appropriate. In each case explain your reasoning.

(i) 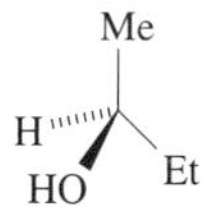 Me H HO Et (ii) 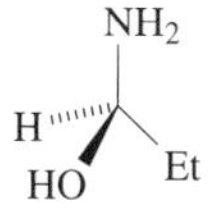 NH_2 H HO Et

(iii) 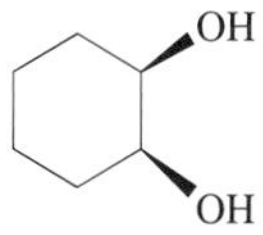 OH OH (iv) 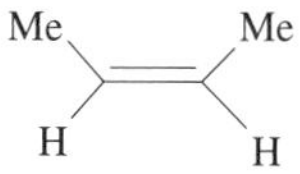 Me Me H H

(v) 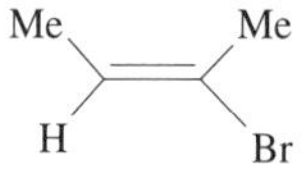 Me Me H Br (vi) 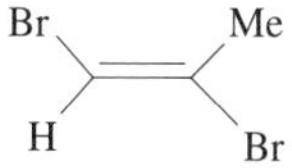 Br Me H Br

99. Give examples to illustrate **FIVE** of the following *in the context of chemical reactions* and explain their significance.

(a) An inductively stabilised carbanion.
(b) A mesomerically (resonance) stabilised carbocation.
(c) A nucleophilic substitution reaction with retention.
(d) A reaction which is zero order with respect to one of its reactants.
(e) A reaction involving a slow (rate-determining) transfer of hydrogen.
(f) A *syn* elimination process leading to an alkene.
(g) A tertiary alkyl halide resistant to nucleophilic substitution.

100. Answer **ALL** parts of this question.

(a) The pK_a of ethanoic acid, CH_3COOH, is 4. For a dilute aqueous solution of ethanoic acid, at what pH is the concentration of CH_3COOH equal to the concentration of CH_3COO^-? State the approximate relative concentrations of CH_3COOH and $CHCOO^-$ at:
 (i) pH 5
 (ii) pH 7

(b) For **EACH** of the following pairs of compounds, indicate clearly with reasoning which is the stronger acid in water.

(a) 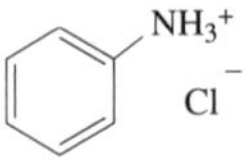 $C_6H_5NH_3^+ Cl^-$ or $C_6H_{11}NH_3^+ Cl^-$

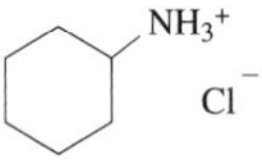

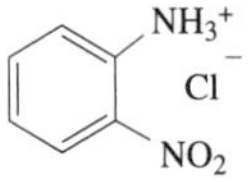

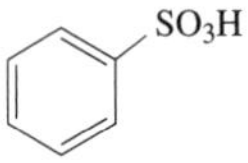

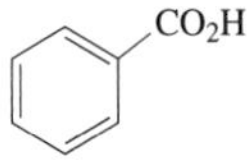

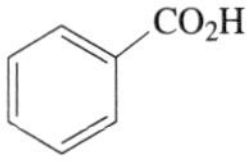

(b) 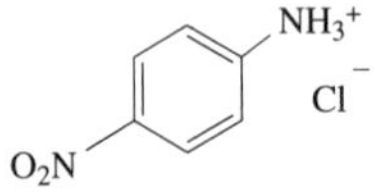 O_2N-$C_6H_4NH_3^+ Cl^-$ (para) or $C_6H_4(NO_2)NH_3^+ Cl^-$ (ortho)

(c) $C_6H_5SO_3H$ or $C_6H_5CO_2H$

(d) $C_6H_5CO_2H$ or $C_6H_{11}CO_2H$

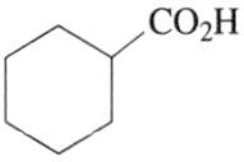

(e) CH_3CONH_2 or $CH_3SO_2NH_2$

101. Write balanced equations for **ALL** of the following reactions indicating whether the overall process involves oxidation, reduction or neither. Where relevant, identify which species are being oxidised and which species are being reduced:

(a) PhC(O)Ph —(i. N_2H_4; ii. $KOBu^t$)→ $PhCH_2Ph$

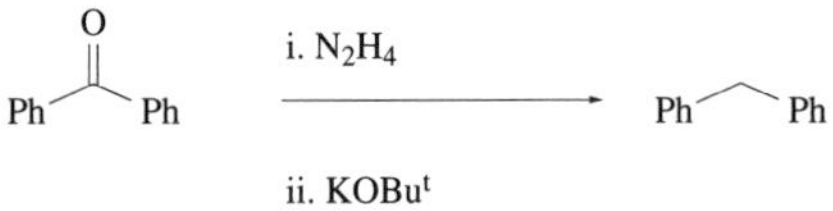

(b) cyclohexene —(i. HOBr, aq. THF; ii. KOH, H_2O, heat)→ cyclohexene oxide (O)

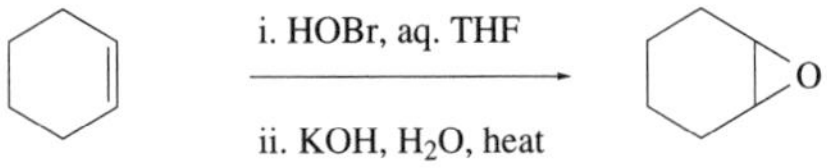

(c) 3,4-(MeO)$_2$C$_6$H$_3$CHO —(i. NH_2OH, H_2O, EtOH; ii. $(CH_3CO)_2O$, heat)→ 3,4-(MeO)$_2$C$_6$H$_3$CN

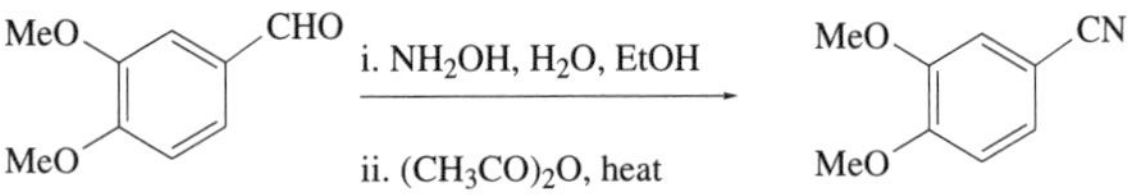

(d) 2-Cl-C$_6$H$_4$CH$_3$ —($KMnO_4$, H_2O, heat)→ 2-Cl-C$_6$H$_4$COOH

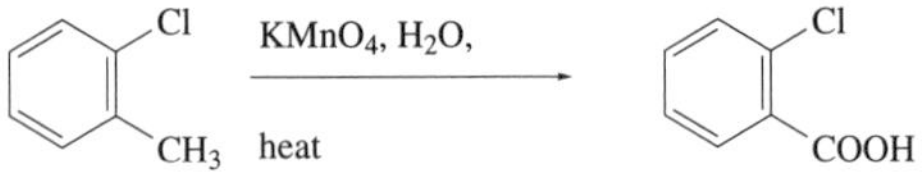

(e) 4-HO-C$_6$H$_4$CF$_3$ —(i. NaOH, H_2O; ii. H_3O^+)→ 4-HO-C$_6$H$_4$COOH

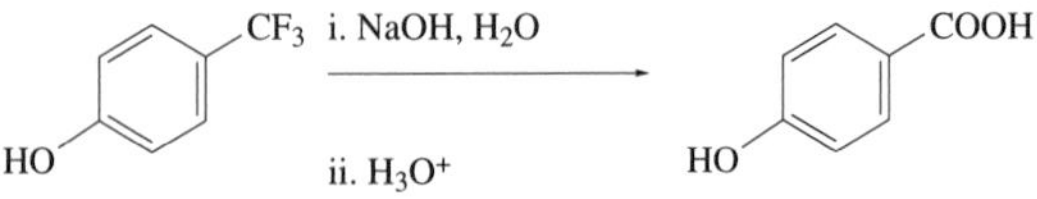

102. Answer **ALL** parts of this question.

(a) How many stereoisomers of 1,2,3-trichlorocyclobutane, below, are there?

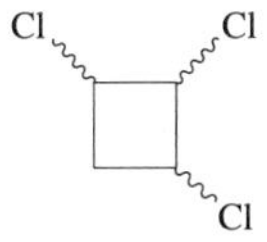

Cl Cl Cl

Draw out three-dimensional representations of them indicating, where relevant, diastereomeric and enantiomeric relationships.

(b) Draw out the three-dimensional representation of (2*S*,3*S*)-3-bromobutan-2-ol [gross structure: MeCHBrCHOHMe]. On treatment of (2*S*,3*S*)-3-bromobutan- 2-ol with HBr, racemic 2,3-dibromobutane is obtained; explain.

What would you expect to be the result of treating (2*S*,3*R*)-3-bromobutan-2-ol with HBr under the same conditions?

103. Suggest suitable reagents for carrying out **ALL** the transformations **A** to **K**; more than one step may be necessary for some of these transformations.

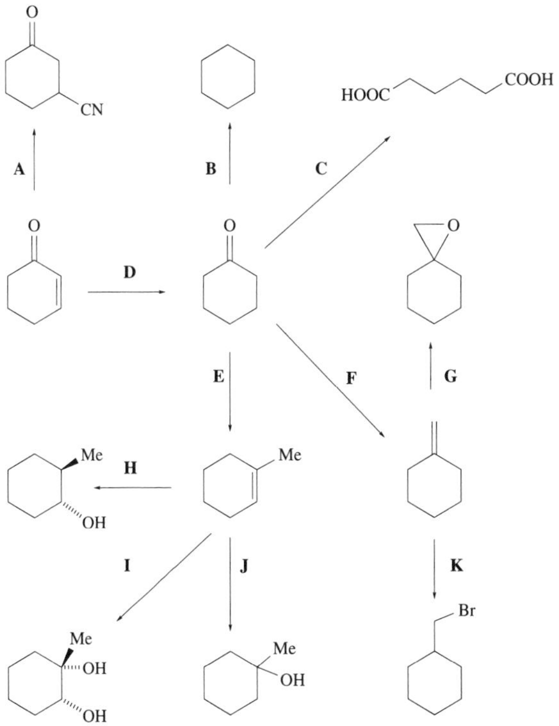

O CN

HOOC COOH

A **B** **C**

O O O

D

E **F** **G**

Me Me

H

OH

I **J** **K**

Br

Me OH Me OH

OH

104. Give the IUPAC names for these compounds:

(i) 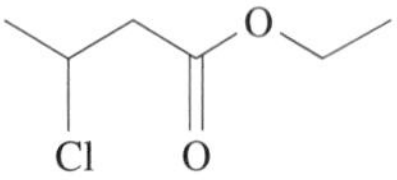 O Cl O (ii) 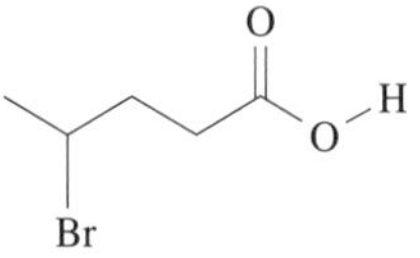 O H O Br

(iii) 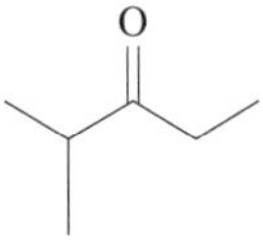 O (iv) 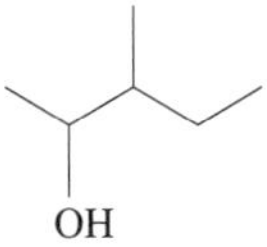 OH

What reaction product(s) would you expect from the following reactions?

(i) 1-propanol with acetic acid (CH_3COOH) in the presence of a trace of strong acid.
(ii) Pentanoic acid with ammonia (NH_3).
(iii) Methylamine with an excess of methyl iodide in the presence of NH_4OH.
(iv) 2-Butanone with methanol in the presence of sulfuric acid.

Outline the reaction mechanisms involved in **TWO** of the reactions.

105. Draw the structures corresponding to these IUPAC names.

(i) 2-Ethyl-2-buten-1-ol.
(ii) 3-Methylbutanal.
(iii) 2,3-Dimethylhexanoic acid.
(iv) 4-Chloro-2-pentanone.

The Grignard reaction is one of the most important reactions used by the chemist in the laboratory. Outline the mechanism of the Grignard reaction. How could you use a Grignard reaction to prepare the following alcohols?

(i) 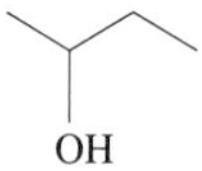 OH (ii) 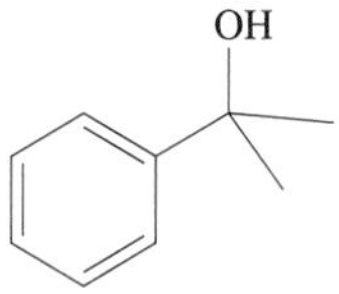 OH

(iii) 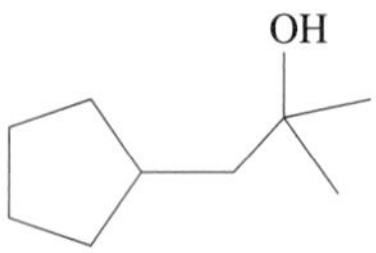 OH (iv) 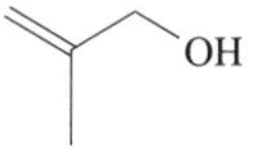 OH

List some limitations of the Grignard reaction.

106. Answer **TWO** parts from (a) to (c).

(a) Account for the basicity of amines mentioning the factors that affect basicity.
Arrange the following in order of increasing basicity.

(i) CH_3NH_2, NH_3, $(CH_3)_2NH$

(ii) $ClCH_2NH_2$, CH_3NH_2, FCH_2NH_2

Give reasons for your answer.

(b) How would you prepare TWO of the following compounds using a Williamson ether synthesis?
(i) Methyl propyl ether.
(ii) Ethyl 2,2-dimethylpropyl ether.
(iii) Ph-O-CH_2CH_3.

(c) Describe a useful synthesis for the four following compounds, starting from benzene. Briefly comment on the reasons for the choice of reaction route that you have made.

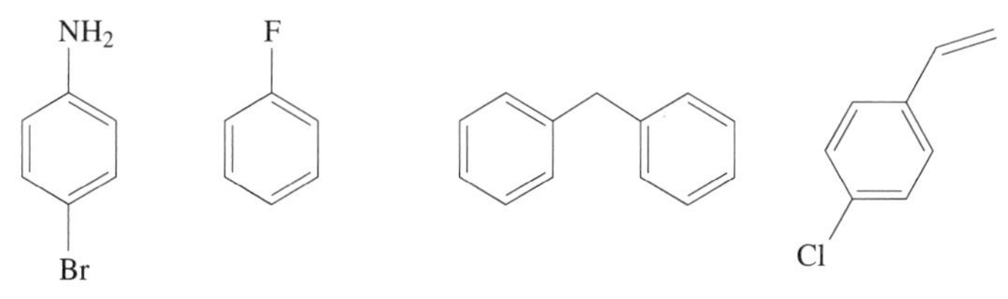

107. Answer **TWO** parts from the following (a) to (d).

(a) Suggest reasons why the nitration of biphenyl **A** occurs mainly at the *ortho* and *para* positions.
At what position and in what ring is *p*-bromobiphenyl **B** likely to be nitrated?

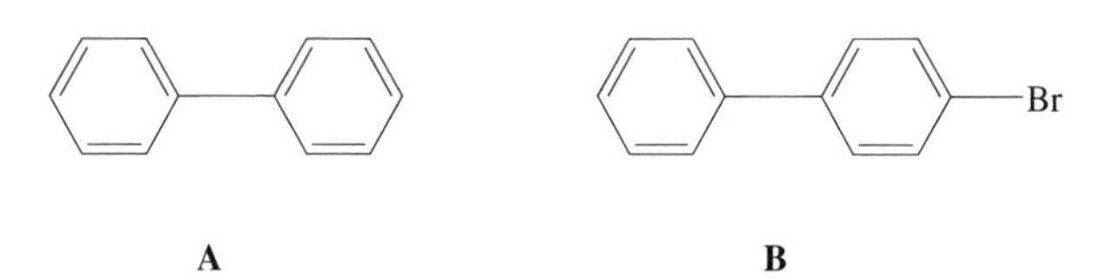

(b) Diazonium salts are useful reagents in synthetic organic chemistry. Illustrate this referring to (i) formation of diazonium salts from amines, (ii) introduction of halogens, (iii) phenol formation, (iv) formation of azo dyes (including pH control of the reaction medium).

(c) In the chemistry of naphthalene and 1,3-dienes the product formed in a reaction may not be the thermodynamically most stable isomer. Show how this can arise referring to (i) reaction conditions, (ii) reaction coordinate diagrams, (iii) isotope effects.

(d) Write a note on the discovery of benzyne, referring to the use of isotopic labelling, trapping reactions and the evidence for its structure.

108. Identify the compounds **A** to **F** for the reaction schemes (a) to (c). You may assume aqueous workups when required. Give reaction mechanisms for all reactions.

A $\xrightarrow{B_2H_6}$ **B** $\xrightarrow[NaOH]{H_2O_2}$ (2-methylcyclohexanol: Me, OH)

C $\xrightarrow{OsO_4}$ **D** $\xrightarrow[Cat.\ H^+]{Me_2CO}$ (Ph, Ph, O, O, Me, Me acetonide)

E $\xrightarrow{MeMgBr}$ **F** $\xrightarrow[(COCl)_2,\ Et_3N]{Me_2S=O}$ PhC(O)Me

109. Show how the following transformations may be carried out. Explain briefly the reasons why the conversions were not best effected in single-stage transformations.

(a) → A 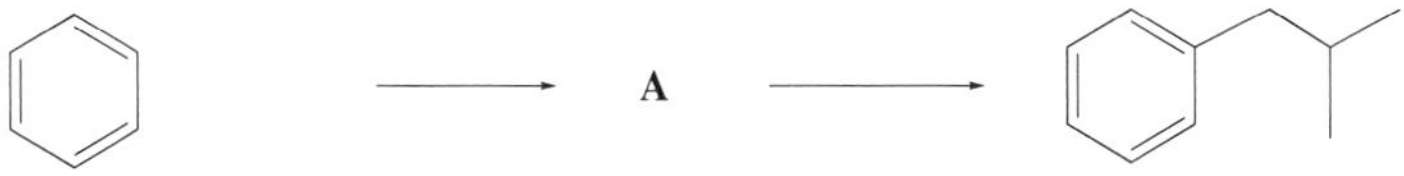 →

(b) 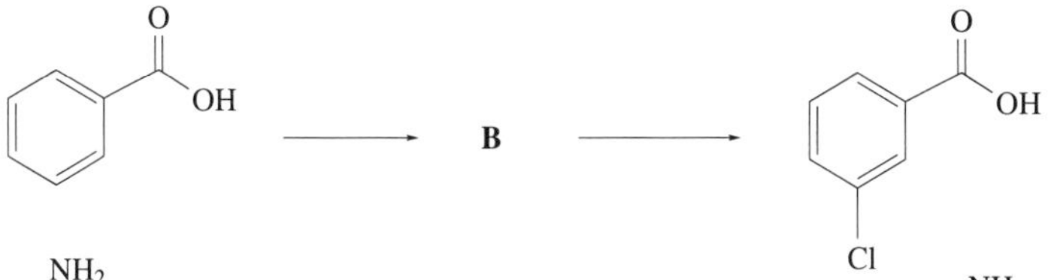 → B →

(c) 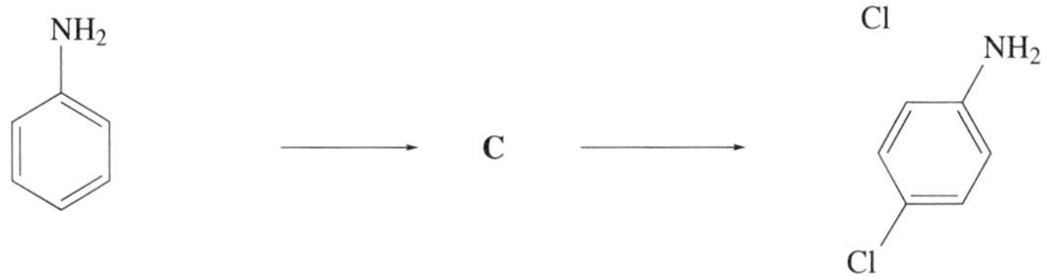 → C →

(d) → D 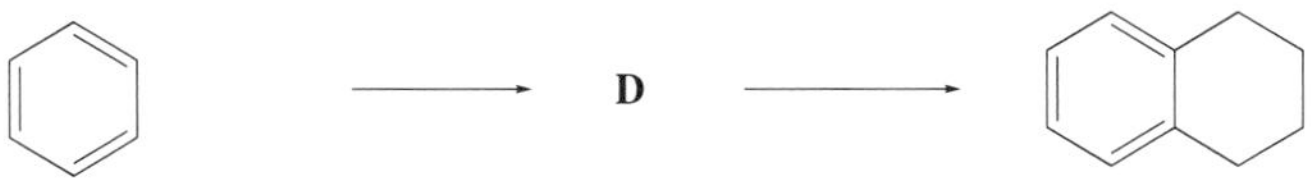 →

110. (a) Give a detailed account of the bonding in ethene.

(b) How might propane, propan-1-ol and 2-bromopropane be prepared from propene? Comment on the mechanisms of the reactions used.

111. (a) Use IUPAC nomenclature to name the following compounds:

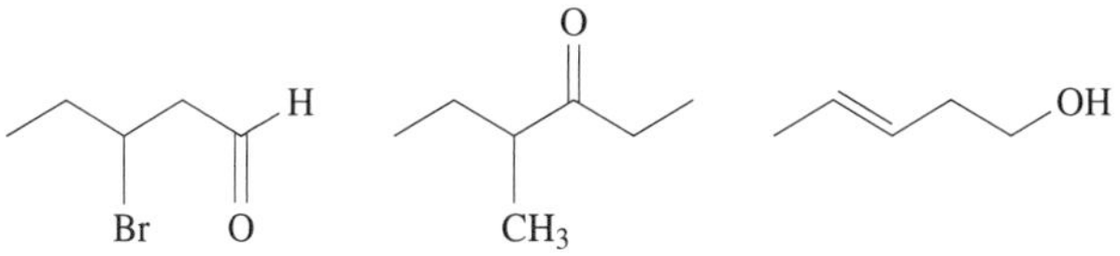

(b) Define the terms nucleophile, electrophile and base. Give an example of each and indicate the type of reactions in which these species participate.

112. (a) Explain how electrophilic substitution of a monosubstituted benzene, C_6H_5Y, is influenced by the nature of the group Y.

(b) Suggest how the following compounds could be made usin-gbenzene as the starting material.

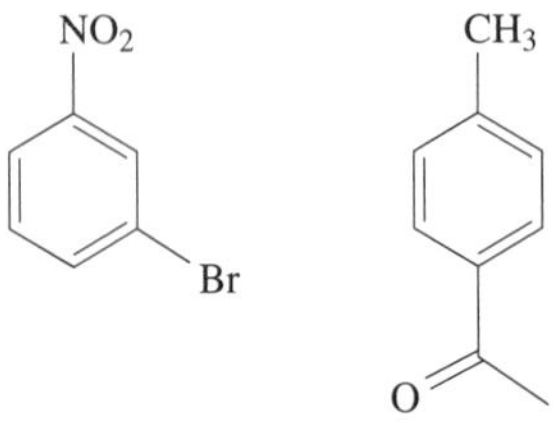

113. What do you understand by "aromatic character". Why does the starting compound in (a) ionise to generate a bromide ion whereas the starting compound in (b) ionises to form a bromonium ion?

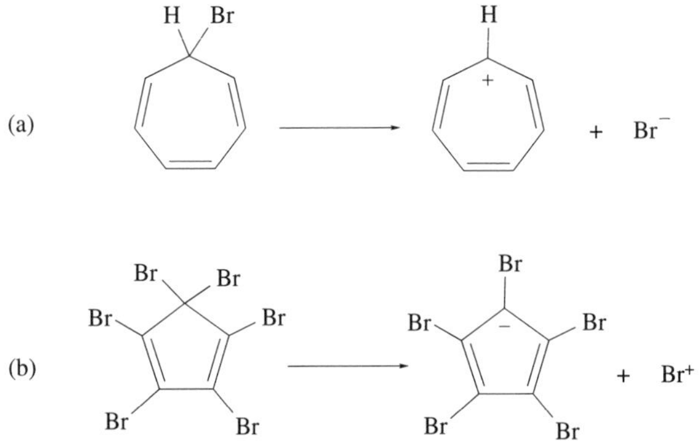

(a)

(b)

Hydrocarbon A is described as an "anti-aromatic" compound. What is meant by this term?

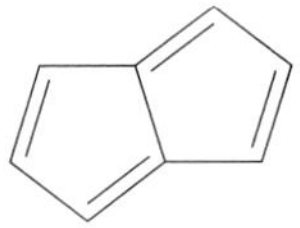

A

114. Answer **BOTH** parts (a) and (b).

(a) Explain, with examples, how infrared spectroscopy may be used to distinguish between inter- and intramolecular hydrogen bonding.

(b) Discuss the mechanism for the mononitration of benzene. Explain why it is much more difficult to carry out the dinitration of benzene and why 1,2- and 1,4-dinitrobenzenes are not formed in the reaction?

How would you convert 1,3-dinitrobenzene into 3-bromoaniline and how would you convert the latter into 3-bromophenol and 3-iodobromobenzene?

115. (a) Describe with the aid of examples how alkyl halides may be prepared from (i) alcohols and (ii) alkenes. Include in your answer a description of the mechanisms involved.

(b) But-1-ene can be converted into a variety of products by the sequence of reactions shown in the following scheme.

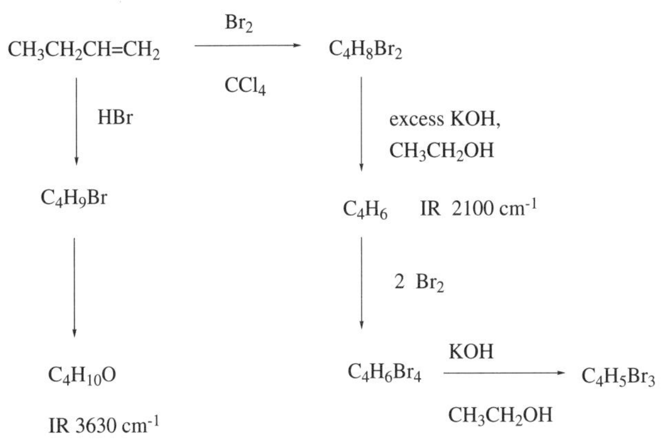

(i) Suggest structures for all of the compounds formed.

(ii) Provide explanations of all the reactions involved.

116. For each of the following reactions draw the structure of and name the product formed.

(a) $CH_3C{\equiv}CH \xrightarrow[\text{ii. } CH_3CH_2I]{\text{i. NaNH}_2\text{, liquid NH}_3}$

(b) $(CH_3)_3CBr \xrightarrow{\text{KOH, H}_2\text{O}}$

(c) $CH_3CH_2CH(OH)CH_3 \xrightarrow{\text{Na}_2\text{Cr}_2\text{O}_7\text{, H}_3\text{O}^+}$

(d) $(CH_3)_2CHBr \xrightarrow{\text{Zn/H}_3\text{O}^+}$

(e) $CH_3CH{=}CH_2 \xrightarrow{\text{HCl}}$

(f) $CH_3CHICH_2CH_3 \xrightarrow{\text{KCN}}$

117. (a) When optically pure (+)-(*S*)-2-chlorobutane is allowed to react with potassium iodide in acetone in an S_N2 as shown below, the 2-iodobutane that is produced has a minus (−) rotation.

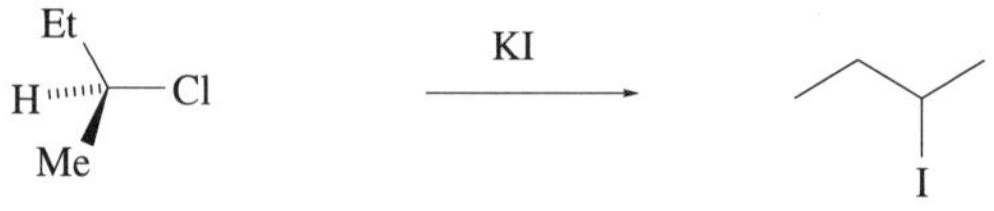

Et
H Cl
Me
KI

I

(+)-(*S*)-2-chlorobutane (-)-2-iodobutane

Draw the configuration of (i) (−)-2-iodobutane and (ii) (+)-2-iodobutane and label each molecule (*R*) or (*S*).

(b) Write the Fischer projection formula for each of the following:

(i) *Meso*-2,3-dibromobutane.

(ii) An optically active 2,3-dibromobutane.

(iii) (*R*)-1,2-Dichloropropane.

For each of the compounds (i) and (ii) draw all three staggered Newman projections.

118. (a)Explain briefly why enolates of carbonyl compounds are more useful as reaction intermediates in alkylation reactions than are enols.

(b) Suggest reagents for each of the following reactions:

(i) CH_3COCH_3 → OH, Me, Me, Et

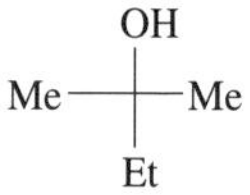

(ii)  CHO →

(iii) $CH_3CH_2CO_2H$ → $CH_3CH_2CH_2OH$

(iv) $CH_2(CO_2CH_3)_2$ → $CH_3CH_2CH(CO_2CH_3)_2$

(v)  CH_2OH → CHO

119. (a) Write a specific example of each of the following reactions and use the curved arrow symbolism to illustrate the mechanism of any **ONE** of the reactions:

(i) Claisen condensation.
(ii) Electrophilic aromatic acylation.
(iii) Cyanohydrin formation.
(iv) Ester hydrolysis under basic conditions.

(b) Write the structural formula of the principal organic product of each of the following reactions:

(i) 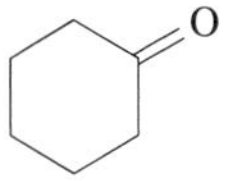 [cyclohexanone] + CH_3OH + HCl ⟶

(ii) 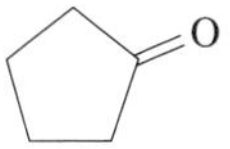 [cyclopentanone] + $NaNH_2$ + CH_3I ⟶

(iii) 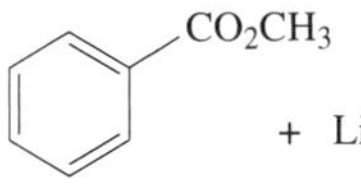 [$C_6H_5CO_2CH_3$] + $LiAlH_4$ followed by H_3O^+ ⟶

(iv) 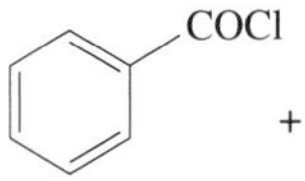 [C_6H_5COCl] + $(CH_3)_2NH$ ⟶

(v) $CH_3COCH_2CH_3$ + $H{-}C{\equiv}C^-\ Na^+$ ⟶

120. (a) Classify the effect of each of the following substituents in electrophilic aromatic substitution according to its ability to:

(i) Increase reactivity.
(ii) Decrease reactivity.
(iii) Lead to *ortho–para* substitution.
(iv) Lead to *meta* substitution.

Substituents: $-NO_2$, $-NH_2$, $-CN$, $-Br$, $NHCOCH_3$, $-OCH_3$, $-OH$, CH_3.

(b) Explain clearly the differences between enantiomers and diastereomers.

Are the following pairs of compounds enantiomers or diastereomers? Explain.

(i) (+)-Tartaric acid and (−)-tartaric acid.
(ii) (−)-Tartaric acid and *meso*-tartaric acid.
(iii) (+)-*Cis* and (−)-*cis*-3-methylcyclohexanol.
(iv) *Cis*- and *trans*-3-methylcyclohexanol.

121. Discuss the mechanism of nitration of aromatic compounds. Account for the products formed by the nitration of toluene, chlorobenzene and nitrobenzene.

122. "Primary aromatic amines are versatile starting materials for the synthesis of other aromatic compounds." Discuss, outlining the mechanisms of any reaction involved.

123. Starting from ethyl acetate or any compound easily made from it, show how the following compounds can be formed.
(a) $CH_3COCH_2CH_2CH_2CH_3$
(b) $CH_3CH_2CH_2CO_2H$
(c) $CH_3CH_2CH_2COCH_2CH_2CH_3$
(d) $CH_3COCH(CH_2CH_3)_2$

124. Explain the meanings of the terms "diastereomer", "enantiomer" and "conformer" (being careful to distinguish the latter from "conformation").

For each of the pairs depicted below, state which of the foregoing terms (if any) may be used to describe the interrelationship, and give a clear explanation of your reasoning.

(a) 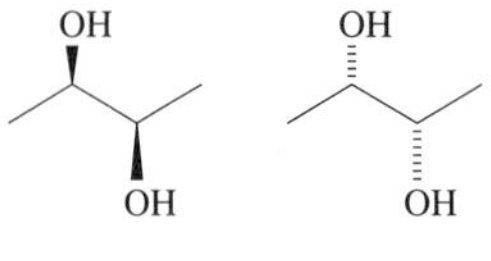 OH OH OH OH (b)

(c) 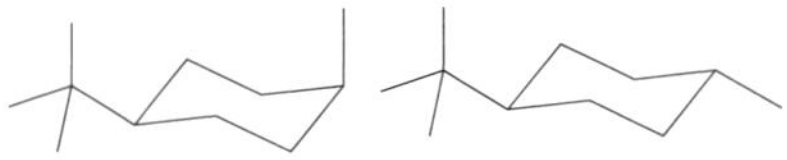 (d)

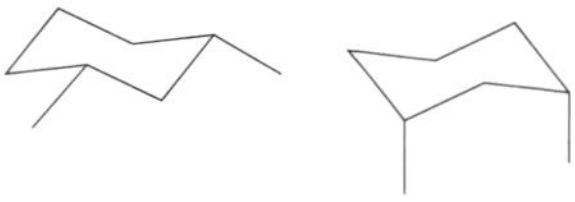

(e) 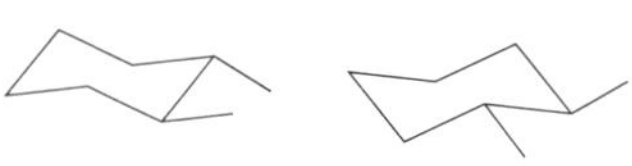 (f) 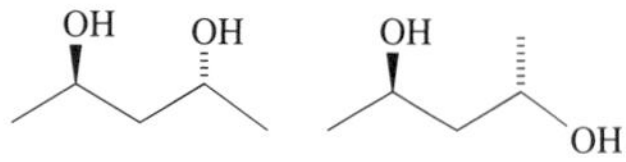 OH OH OH OH

(g) 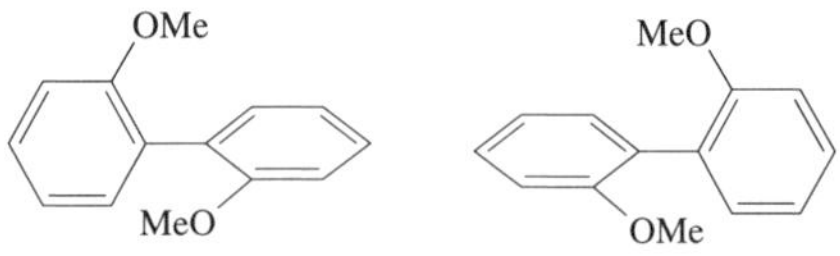 OMe MeO MeO OMe

125. (a) Reduction of **1** with Li(s-Bu)$_3$BH [s-Bu = secondary butyl, $CH_3CH_2CH(CH_3)$-] gives a mixture of **2** and **3** in which **2** predominates. If **2** and **3** are separated and then acetylated with acetic anhydride in pyridine, it is found that **3** reacts much more rapidly than **2**. Provide explanations for the foregoing results.

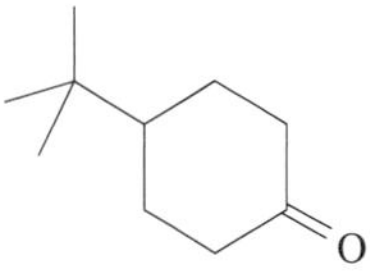

O OH OH

1 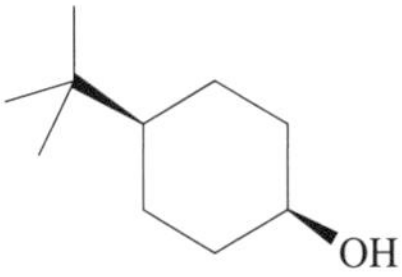 **2** 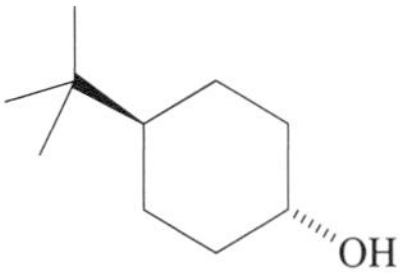 **3**

(b) Explain briefly why IR spectroscopy is of value in identifying the functional groups in a molecule, but is less useful for determining the connectivity of the carbon framework.

126. For **TWO** of the following reactions, give a detailed mechanism for the formation of product.

For the two reactions you have chosen, give alternative preparations of the products, not necessarily using the same starting materials.

(a) NH_2 —CH_2O / HCO_2H→ N(Me)Me

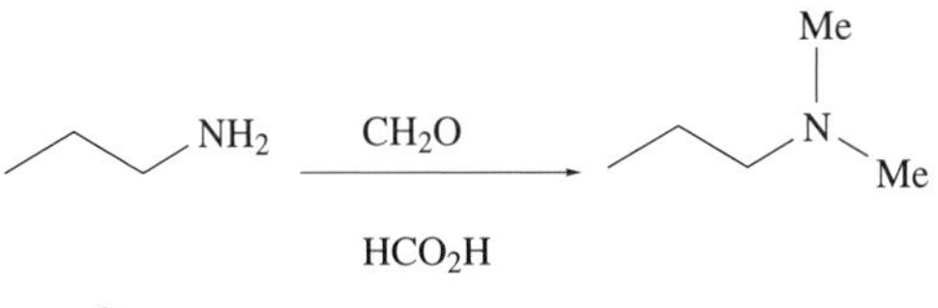

(b) Ph(C=O)Me —N_2H_4, KOH / heat→ $PhCH_2Me$ (H, H)

(c) cyclohexanol (OH) —Me_2SO / $(COCl)_2$ / Et_3N→ cyclohexanone (O)

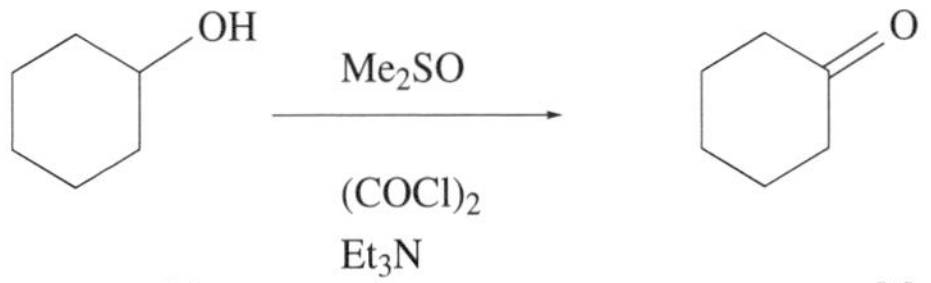

(d) Me, BR_2 (cyclopentane) —H_2O_2/HO^-→ Me, OH (cyclopentane)

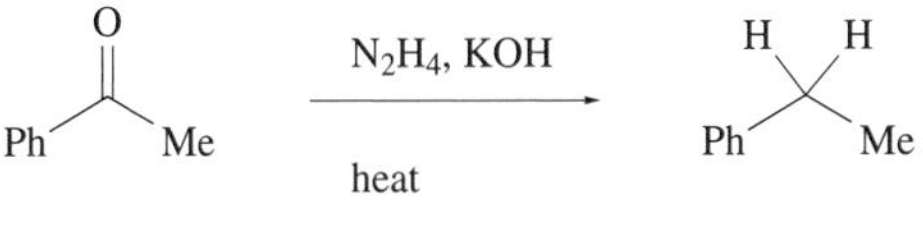

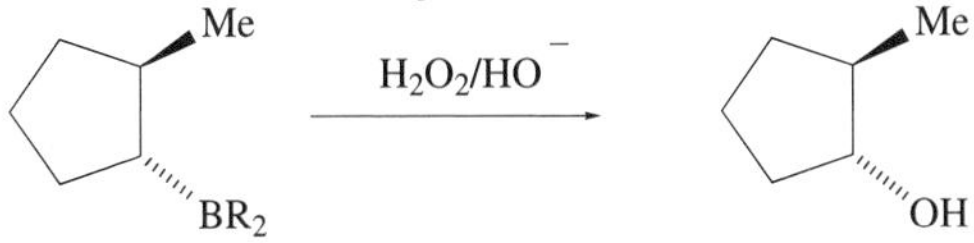

127. Show how the following transformations may be carried out. Explain briefly the reasons why the transformations require more than one step.

(a) [1,3-benzodioxole] ⟶ **A** ⟶ [1,3-benzodioxole-5-CHO]

(b) [4-methylaniline (Me, NH_2)] ⟶ **B** ⟶ [2-bromo-4-methylaniline (Me, NH_2, Br)]

(c) [PhCOCH_2CH_2COOH (O, O, OH)] ⟶ **C** ⟶ [1-tetralone (O)]

(d) [PhCOCH_2NHCOPh (O, NH, O, Ph)] ⟶ **D** ⟶ [1-phenylisoquinoline (N, Ph)]

128. (a) Why do carbon–carbon **triple** bonds vibrate at higher frequencies than carbon–carbon **double** bonds? Relative to these, where might you expect to find absorption bands for C–Cl and C–Br.

(b) Each pair of isomeric compounds below can be easily distinguished by their infrared spectra. Draw rough sketches, labelled with approximate vibration frequencies, to show the spectrum you would expect for each compound.

(i) $CONH_2$ and CN HO

(ii) $CH_3CH{=}CHCH_2OH$ and $CH_3CH_2COCH_3$

129. (a) What categories of organic molecule show strong absorption in the ultraviolet/visible region of the electromagnetic spectrum?

(b) Which of the following isomeric compounds would show the longest wavelength electronic absorption, and why?

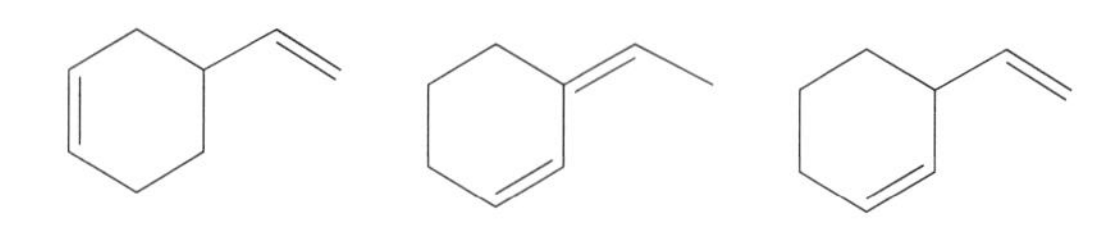

(c) Suggest a reason why the λ_{max} for *trans*-stilbene (PhCH═CHPh) is 295 nm while that for *cis*-stilbene is only 280 nm.

130. Describe the relationship between each pair of structural formulae below as that of *structural* isomers, *enantiomers*, two drawings of the *same* compound (possibly in different conformations), *diastereomers*, or *different* compounds that are not isomers. In each case explain your reasoning.

(i)

(ii)

(iii)

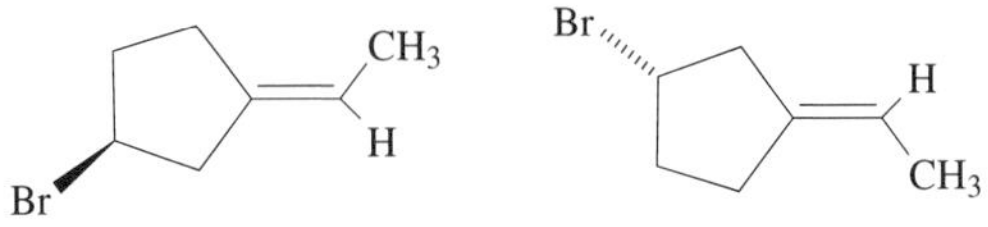

(iv)

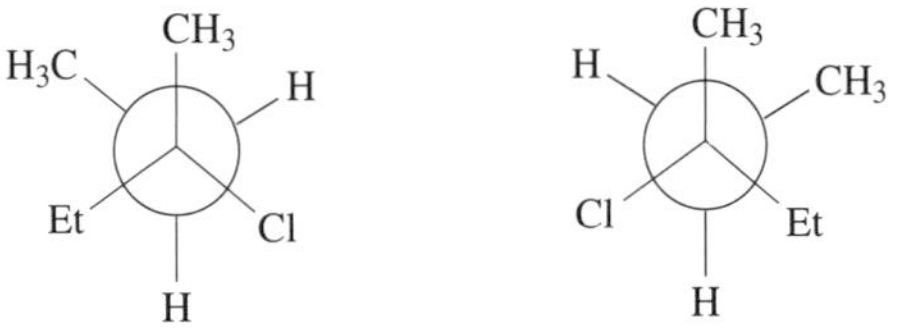

(v)

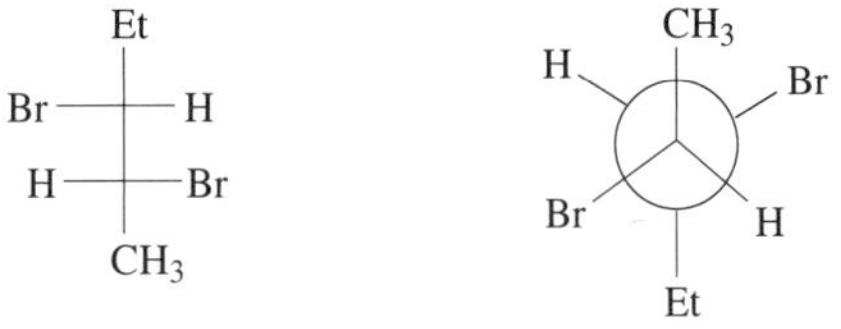

(vi)

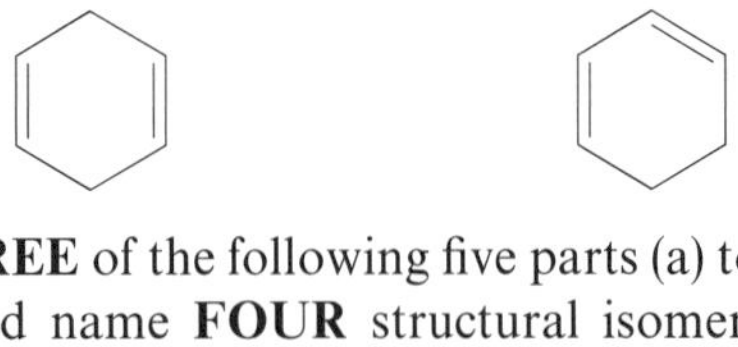

131. Answer **THREE** of the following five parts (a) to (e).

(a) Draw and name **FOUR** structural isomers which have the molecular formula C_3H_5Cl. Pick one of these, show the polarity of the carbon–chlorine bond and explain why it is so.

(b) Define the terms enantiomer, diastereomer and meso compound. Draw both enantiomers of $CH_3CH(OH)CH_2CH_3$ and use the Cahn–Ingold–Prelog rules to assign the configuration (*R* or *S*) of one of them. How many stereoisomers are there of the structure $CH_3CHBrCHBrCH_3$?

(c) Compare and contrast the S_N1 and S_N2 reactions under the following headings:
  (i) Kinetics.
  (ii) Stereochemistry.
  (iii) An example of a molecule which undergoes the reaction.
  (iv) Reaction coordinate diagram.

(d) Four different cycloalkenes can result in the formation of methylcyclopentane if treated with hydrogen over a suitable catalyst. Give the structures of these alkenes and give a suitable catalyst. Draw the product of the reaction of bromine with one of these alkenes paying attention to stereochemistry.

(e) Identify the organic products, compounds **A** to **E** in the following reactions:

(i) $CH_3CH_2CH_2OH + PBr_3 \longrightarrow$ **A**

(ii) $CH_3CH_2CH_2MgBr + H_2O \longrightarrow$ **B**

(iii) $CH_3CH_2CH_2Cl + (CH_3)_2CuLi \longrightarrow$ **C**

(iv) cyclohexene $+ CH_3CO_3H \longrightarrow$ **D**

(v) $Ph{-}C{\equiv}C{-}Ph \xrightarrow{H_2/Pd/CaCO_3/Pb(OAc)_2}$ **E**

132. *Trans*-4-methylcyclohexanol **A** can be converted into compounds **B** to **F** by the sequence of reactions shown below. Identify the compounds **B** to **F**, and comment briefly on reaction mechanisms involved. Pay particular attention to the stereochemistry of reactions (where relevant), and also assign the IR absorptions that are given.

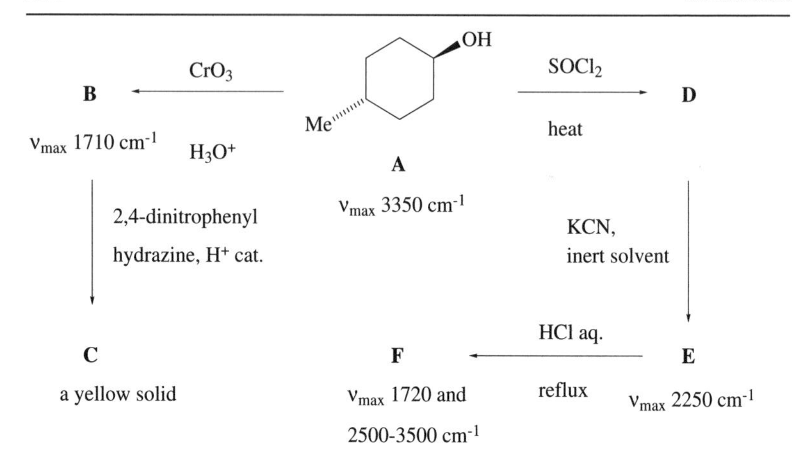

133. What products would you expect from reaction of the *R*-enantiomer of 2-bromobutane with each of the following reagents (you can assume that all reactions are carried out in suitable inert solvents):

(a) Sodium cyanide.
(b) Potassium *tert*-butoxide ($Me_3CO^- K^+$).
(c) Sodium iodide, followed by sodium cyanide.
(d) Sodium ethanoate ($MeCO_2^- Na^+$), followed by aqueous sodium hydroxide.
(e) Magnesium metal, followed by benzaldehyde.

134. Suggest methods and reagents for carrying out the following transformations:

(a) Cyclohexanone to cyclohexane.
(b) Benzonitrile (PhCN) to acetophenone (PhCOMe).
(c) Butanoyl chloride ($MeCH_2CH_2COCl$) to butylamine ($MeCH_2CH_2CH_2NH_2$).
(d) Phenylethanal ($PhCH_2CHO$) or 1,1-dimethoxy-2-phenylethane.
(e) Cycloheptanone to 1-methylcyclohept-1-ene (two steps required).

135. Complete the following reactions:

(a) [cyclohexene] $\xrightarrow{Br_2}$

(b) [cyclooctene] $\xrightarrow[\text{ii. } NaBH_4\text{, } NaOH]{\text{i. } Hg(OAc)_2}$

(c) Me [1-methylcyclopentene] $\xrightarrow[\text{ii. } H_2O_2\text{, } NaOH]{\text{i. } BH_3\text{, THF}}$

(d) Me–CH=CH–Me $\xrightarrow[\text{ii. } H_2O\text{, } H_2SO_4]{\text{i. MCPBA}}$

(e) [1-methylcyclohexene] Me $\xrightarrow[\text{ii. } Me_2S]{\text{i. } O_3}$

(f) H—≡—Me $\xrightarrow[\text{ii. cyclopentanone; iii. } H_2O\text{, } H_2SO_4]{\text{i. } NaNH_2/NH_3 \text{ (liq.)}}$

136. Predict the products of the following reactions. Explain why the product is formed and elaborate probable mechanisms.

(a) [benzene ring]–CO_2Me $\xrightarrow{\text{conc. } H_2SO_4/HNO_3}$

(b) Cl–[benzene ring]–NH_2 (para) $\xrightarrow{NaNO_2/\text{aq. HCl, 0 °C}}$

(c) [benzene ring with F, NO_2 (ortho to F) and NO_2 (para to F)] $\xrightarrow{Pr_2NH}$

137. Answer any **THREE** parts from (a) to (e).

(a) Four different cycloalkenes yield methylcyclopentane when subjected to catalytic hydrogenation. What are their structures?

Taking one of the cycloalkenes as an example, show the product obtained on bromination (taking particular care to show the stereochemistry).

(b) For the following reactions, supply the missing reagents or product, briefly explaining the reason for your choice:

(i) [benzene ring with CO_2Me] $\xrightarrow{\text{conc. } H_2SO_4/HNO_3}$

(ii) [benzene ring with NH_2 and Cl (para)] $\xrightarrow{NaNO_2\text{/aq. HCl, 0 °C}}$

(iii) [benzene ring with F, NO_2 (ortho) and NO_2 (para)] $\xrightarrow{Pr_2NH}$

(c) Comment on the following order of reactivity shown for the reaction of sodium hydroxide with alkyl halides (by S_N2 mechanism):

$CH_3Br > CH_3CH_2Br > (CH_3)_2CHBr > (CH_3)_3CCH_2Br > (CH_3)_3CBr$

Most reactive Least reactive

(d) Compare the conditions required for acid catalysed dehydration of primary, secondary and tertiary alcohols to alkenes (with suitable examples). What intermediate is formed in these reactions?

Some alkenes have a rearranged structure when formed in this way–how does this arise?

(e) Define and give a suitable example of the following:

(i) An enantiomer.
(ii) A meso compound.
(iii) A bicycloalkane.
(iv) Resolution of enantiomers.

138. (a) Explain precisely the terms "constitution" and "configuration" as used to describe isomers. Do the following pairs of compounds differ from each other in constitution or configuration? Explain.

(i) Lactic acid ($CH_3CH(OH)COOH$) and β-hydroxyp-

ropanoic acid ($HOCH_2CH_2COOH$).
(ii) (+)-Lactic acid and (−)-lactic acid.
(iii) 3-Methylcyclohexanol and 4-methylcyclohexanol.
(iv) *Cis*- and *trans*-4-methylcyclohexanol.
(v) 1-Chloropropene and 3-chloropropene.

(b) Draw structural formulae for the isolable stereoisomers of the following compounds and indicate the type of isomerism.

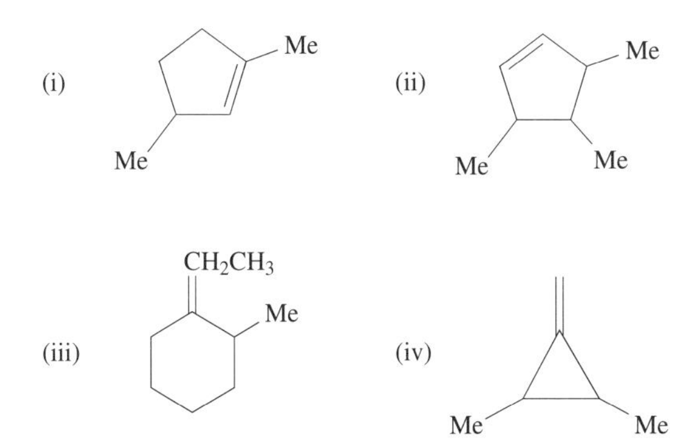

139. (a) Give an account of the methods available for the synthesis of carboxylic acids and esters.
(b) Explain in detail why ester formation is acid catalysed but not base-catalysed.

140. Provide mechanistic rationalisms for any **FOUR** of the following reactions:

(i) O $\xrightarrow{NH_2OH}$ N OH

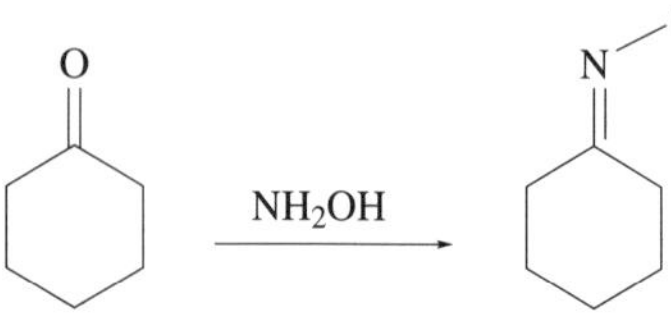

(ii) O $\xrightarrow[\text{then acid}]{MeMgBr}$ OH

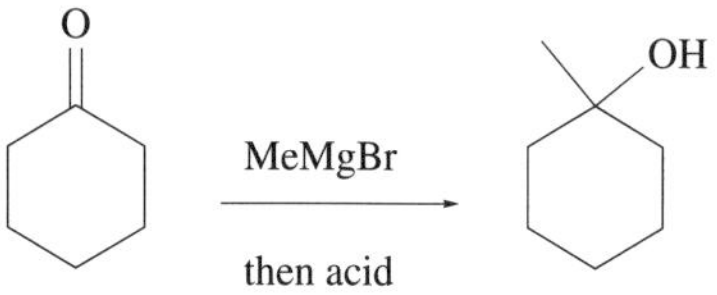

(iii) Ph NH_2 O $\xrightarrow{NaOH/Br_2}$ Ph NH_2

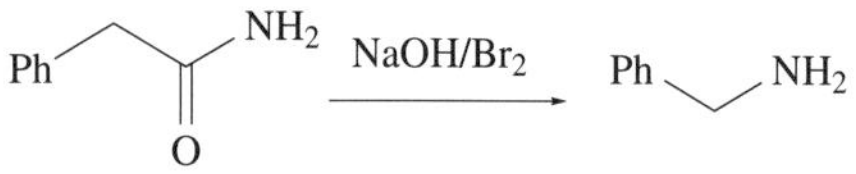

(iv) Br $\xrightarrow{NaCN}$ CN

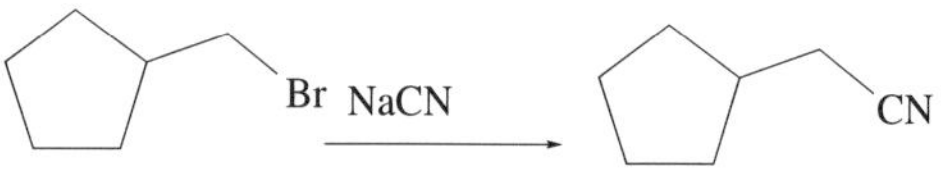

(v) Ph Br $\xrightarrow{{}^{t}BuO^{-}\ K^{+}}$ Ph

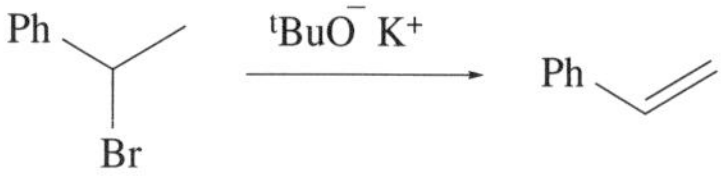

(vi) Br $\xrightarrow{H_2O}$ HO

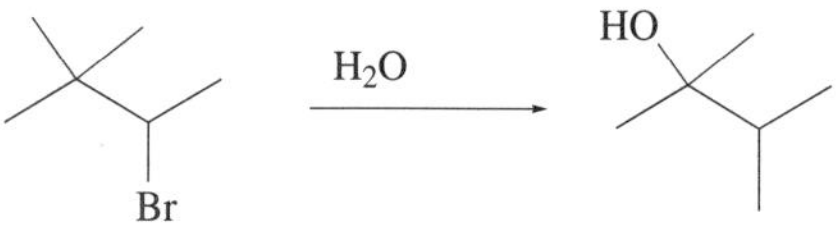

141. Explain the following observations:

(a) Treatment of benzene with 1-chloro-2-methylpropane and aluminium trichloride gives mostly 2-methyl-2-phenyl propane.

(b) Bromobenzene undergoes electrophilic substitution more slowly than benzene, even though substitution occurs preferentially in the *ortho* and *para* positions.

142. Answer **TWO** parts.

(a) Name the following six compounds by the IUPAC system.

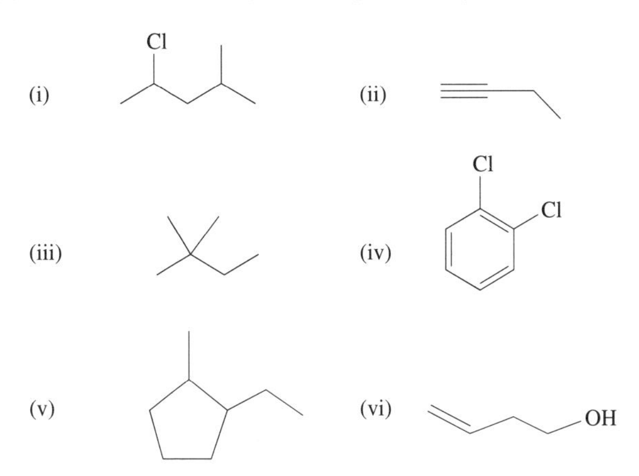

(b) Give the structural formula for each of the following six compounds:
  (i) 3-Methylpentane.
  (ii) 1,3-Butadiene.
  (iii) 3-Butylcyclopentene.
  (iv) Vinyl chloride.
  (v) *Cis*-1,2-dichloroethene.
  (vi) 3,5-Dibromonitrobenzene.

(c) Describe or define each of the following six terms:
  (i) Isotope.
  (ii) Covalent bond.
  (iii) Configuration.
  (iv) sp^2 hybridisation.
  (v) Chiral centre.
  (vi) Electrophilic addition.

143. Answer **TWO** parts.

(a) State Markovnikov's rule and use it to predict the likely outcome of electrophilic addition of HCl to propene.

(b) Describe, using molecular diagrams, the substitution of Cl into a benzene ring, and particularly the role of the catalyst.

(c) Describe the preparation of a Grignard reagent and how it can be used to synthesise a secondary alcohol.

144. Answer **ALL** parts of this question.

(a) The pK_a values for a series of substituted ethanoic acids,

(X)CH(Y)COOH, are: 2.5, 2.8, 4.0 and 4.9. The structures of the acids are given below. By considering the effect of the substituent X, state which pK_a corresponds to which acid.

$CH_3CHClCOOH$ ($X = Cl$, $Y = CH_3$)
FCH_2COOH ($X = F$, $Y = H$)
$CH_3CH_2CH_2COOH$ ($X = CH_2CH_3$, $Y = H$)
$ClCHCHCOOH$ ($X = ClCH_2$, $Y = H$)
Explain your reasoning.

(b) For each of the following pairs of anions, which is the stronger base?

$CH_3CO_2^-$ or $ClCH_2CO_2^-$
$CH_3CO_2^-$ or EtO^-
Cl^- or HO^-
Clearly explain your reasoning.

(c) Why is pyridine a weaker base than piperidine? (Hint: Consider the hybridisation of the nitrogen atom.)

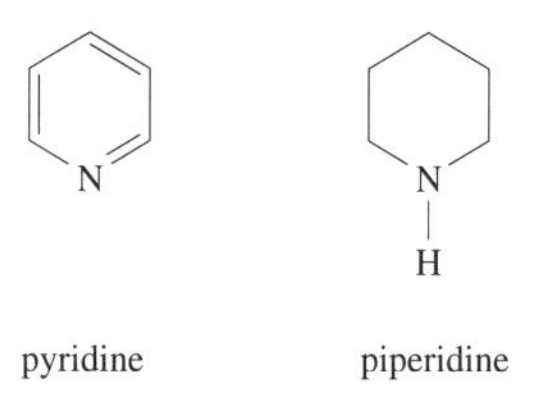

145. Draw the two possible diastereomers of 1-bromo-4-methylcyclohexane. For each isomer draw two possible chair conformations, and in each case, state which conformation has the lower energy.

Trans-2-methylcyclopentanol undergoes the reactions shown below. Suggest structures for the products **B** to **E**. You should clearly explain the mechanisms of the reactions involved, and pay particular attention to the stereochemistry where relevant.

A (OH, Me) —$SOCl_2$, heat→ **B** —NaI, acetone (solvent)→ **C**

A —TsCl, pyridine→ **D** —KCN, DMF (solvent)→ **E**

146. Predict the products of the following reactions. In each case briefly discuss the reaction mechanism involved.

$MeC(=O)CH_2Me$ $\xrightarrow[\text{cat. } H^+]{EtSH}$

PhCHO $\xrightarrow{HO^-}$

Cyclopentanone $\xrightarrow[\text{cat. } H^+]{PhNHNH_2}$

Cyclohexanone $\xrightarrow[\text{cat. } H^+]{Et_2NH}$

$PhC(=O)CH_2Me$ $\xrightarrow[EtOH]{NaBH_4}$

147. Compare and contrast nucleophilic substitution reactions of halo compounds which occur *via* the S_N1 or S_N2 mechanism. To gain full marks you must include a detailed description of factors such as; the nature of the starting halide, the nature of the nucleophile, solvent, and stereochemistry.

148. Suggest methods for carrying out the following transformations.
(a) Cyclohexanone to cyclohexane.
(b) Cyclohexanol to chlorocyclohexane.
(c) Bromocyclohexane to cyclohexene.
(d) Iodocyclohexane to cyclohexanol.
(e) Cyclohexanone to 1,1-dimethoxycyclohexane.

149. Discuss the chemistry of alcohols and include in your discussion a summary of methods for their preparation.
[Your answer should include consideration of the origins of their reactivity, the types of reagent (both inorganic and organic) with which they undergo reaction, specific examples and significant features of their reactions with a variety of these reagents and the mechanisms of these reactions.]

150. Compounds **A** and **B** are isomeric hydrocarbons. On ozonolysis, each yielded the same two isomeric compounds, **C** and **D**. Reaction of **C** with aqueous potassium permanganate yielded an acidic compound **E**. In contrast, compound **D** did not react with potassium permanganate.

Quantitative combustion of 0.021 63 g of **C** in a stream of oxygen yielded 0.052 27 g of carbon dioxide and 0.021 27 g of water.

Reaction of methylmagnesium iodide (MeMgI) with propanol (C_3H_6O) yielded a product **F** which, on reaction with aqueous potassium permanganate, was converted to **D**.

(a) Calculate the molecular formula of **C**.

(b) Outlining your reasoning clearly, assign a possible structure to each of compounds **A** to **F**.

(c) Account for the isomerism of compounds **A** and **B**.

151. A typical free radical substitution usually has three steps: initiation, propagation and termination. Explain this statement with reference to the monochlorination of propane.

$$C_3H_8 + Cl_2 \longrightarrow \text{products}$$

152. The analysis of organic compounds for constituent elements is a very important ' technique in organic chemistry.

(i) Show how the most commonly encountered elements in organic compounds (C, H, N) are quantitatively determined.

(ii) The analysis of a compônd showed that it contained 52.14% carbon, 13.04% hydrogen and 34.80% oxygen. Given that the molecular mass is 46, deduce the molecular formula and draw two structures which correspond to the formula.

153. Answer parts (i) to (iii).

(i) Draw the stereoisomers of the following compounds.

(a)

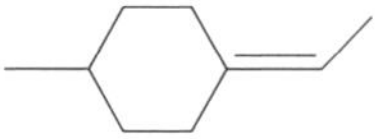

(b) $CH_3CH{=}C{=}CHCH_3$

(c) 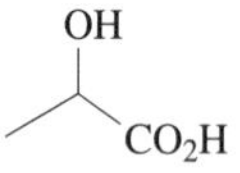 OH, CO_2H

(d) 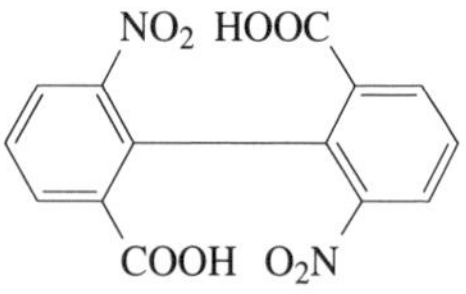 NO_2 HOOC, COOH O_2N

(ii) Draw the stereoisomers of the following compound.

NOH

Use the stereoisomers to illustrate the meaning of the term "diastereomer".

(iii) Identify the products obtained on treating the following componds with bromine.

154. Discuss the chemistry of the aldehydes.

[Your answer should include a discussion of their bonding, the origins of their reactivity, the types of reagent with which they undergo reaction, examples of their reactions with a range of these reagents, comparison of their reactivity with that of ketones, and the role of acids in catalysing their reactions.]

(a) O; Ph Ph — i. N_2H_4; ii. $KOBu^t$ → Ph Ph

(b) i. HOBr, aq. THF; ii. KOH, H_2O, heat → O

(c) MeO, MeO, CHO — i. NH_2OH, H_2O, EtOH; ii. $(CH_3CO)_2O$, heat → MeO, MeO, CN

(d) Cl, CH_3 — $KMnO_4$, H_2O; heat → Cl, COOH

(e) HO, CF_3 — i. NaOH, H_2O; ii. H_3O^+ → HO, COOH

155. Compare and contrast the chemistry of acetic anhydride (MeCOOCOMe) and ethyl acetate (MeCOOEt). Include in your discussion a summary of methods for their preparation from acetic acid.

[Your answer should include discussion of the origins of their reactivity, the types of reagent with which they undergo reaction, examples of their reactions with a range of these reagents, comparison of their reactivity, and the role of acids and bases in catalysing the reactions of ethyl acetate.]

156. An alkyl bromide **A** reacted with magnesium to give a Grignard reagent which, when treated with water, yielded a compound **B** whose elemental composition by mass was C: 82.66% and H: 17.34%.

Treatment of compound **A** with aqueous sodium hydroxide solution yielded four products **C, D, E** and **F**.

Compound **C** reacted vigorously when cautiously added to a pellet of sodium metal, whereas **D, E** and **F** did not react with sodium. When heated with concentrated sulfuric acid, compound **C** was converted to a mixture of **D, E** and **F**.

Reaction of **D, E** and **F** with gaseous hydrogen in the presence of a platinum catalyst yielded compound **B**.

(a) Calculate the molecular formula of B.

(b) Outlining your reasoning clearly, assign a possible structure for compounds A to E.

157. Predict the products for **EACH** of the following reactions. Give clear reasoning for your answers.

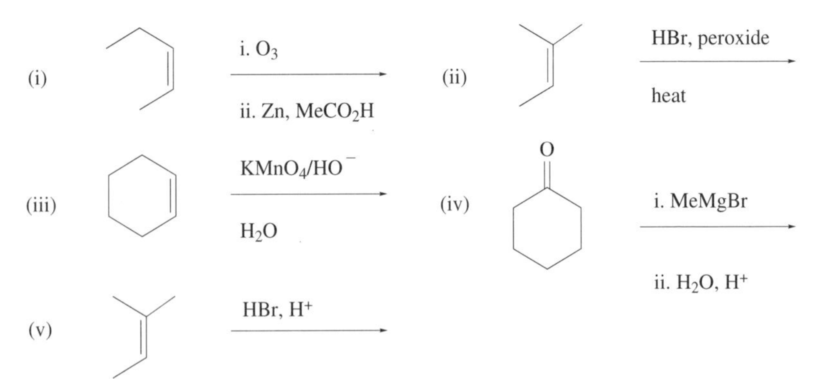

158. Discuss the chemistry of the alkenes and include in your discussion a summary of methods for their preparation.

[Your answer should include consideration of the bonding in alkenes, their structural features, the origins of their reactivity, the types of reagent with which they undergo reaction, specific examples and significant features of their reactions with a variety of these reagents and their mechanisms.]

159. Answer **BOTH** parts.

(i) Show how the elements carbon and hydrogen are quantitatively determined in organic chemistry.

(ii) The antibiotic, penicillin G, gave the following results on analysis:

C 57.45%; H 5.40%; N 8.40%; S 9.60%

The molecular weight is 330 ± 10.

Calculate the molecular formula. (You may assume there are no other additional elements other than oxygen present.)

160. Answer **BOTH** parts.

(a) Write an essay on electrophilic aromatic substitution reactions of benzene.

(b) Discuss, in mechanistic terms, the chemistry involved in **THREE** of the following conversions:

(i) $PhCHO + CH_3COPh \xrightarrow{\text{NaOEt in EtOH}} PhCH{=}CHCOPh$

(ii) $CH_2(CO_2Et)_2 \xrightarrow[\text{ii. } PhCH_2Cl]{\text{i. NaOEt in EtOH}} PhCH_2CH(CO_2Et)_2$

(iii) $PhCH_2CH(CO_2Et)_2 \xrightarrow[\text{heat}]{\text{dil. } H_2SO_4} PhCH_2CH_2CO_2H$

(iv) $PhCH{=}CHCOPh + PhOH \xrightarrow{\text{aq. NaOH}} PhCHCH_2COPh$

161. Answer **ALL** parts.

(a) Write a brief account of keto–enol tautomerism.

(b) Account for the fact that reaction of aniline with fuming nitric acid and fuming sulfuric acid at 140 °C yields *m*-nitroaniline.

(c) How might **TWO** of the following conversions be achieved? More than one reaction step is involved in each case.

(i) Benzene ⟶ $PhNHCOCH_3$

(ii) Toluene ⟶ PhCOOPh

(iii) Aniline ⟶ $Me_2N-C_6H_4-N{=}N-Ph$

(iv) Phenol ⟶ o-$MeCOO-C_6H_4-CO_2H$

162 Answer **ALL** parts.

Butan-2-ol **A** is converted to 2-iodobutane **B** on treatment with hydrogen iodide under appropriate conditions. The rate of formation of **B** is dependent on the concentrations of both **A** and hydrogen iodide.

(a) Does this conversion of **A** to **B** involve an S_N1 or an S_N2 mechanism? Briefly justify your answer.

(b) Account for the obserwation that the alcohol **A** does not react with sodium iodide in the absence of a source of protons.

(c) Wrie a detailed mechanistic reaction sequence for the conversion of **A** to **B**, consistent with your answers to parts (a) and (b).

(d) Using the mechanism detailed in part (c), predict whether the alkyl iodide **B**, obtained by reaction of a pure sample of the *R*-isomer of alcohol **A** with hydrogen iodide, would be expected to (i) have the *R*-configuration, (ii) have the *S*-configuration, or (iii) be racemic. Justify your prediction.

(e) When an optically active sample of the alkyl iodide **B** is stirred for some time in a solution containing radioactive so ium iodide (*i.e.* sodium iodide having the radioactive isotope ^{131}I rather than the more common non-radioactive isotope ^{127}I) and then recovered from te solution and repurified, the recovered alkyl iodide **B** is found o be:

(i) Radioactive.

(ii) Optically inactive.

Account for these two observations.

163. Describe in detail the reactions of aromatic amines in terms of their:

(i) Role as bases.

(ii) Role as nucleophyles.
(iii) Ring substitution.
(iv) Diazonium salt formation.

164. Compare and contrast the S_N1 and S_N2 reactions in the following terms:
(i) Reaction rates.
(ii) Role of substituents.
(iii) Effect of solvent.
(iv) Optical activity.

165. Discuss ring strain in cycloalkanes, including in your answer reference to:
(i) Baeyer strain theory.
(ii) Heats of combustion.
(iii) Newman projections.
(iv) Conformational analysis.

166. Discuss the common electrophilic aromatic substitution reactions of benzene, anisole (PhOMe), and nitrobenzene.
[Your answer should include, but not necessarily be confined to, consideration of the common electrophiles and their generation, the mechanisms by which electrophiles react with aromatic compounds, the relative reactivities of the above three compounds to electrophilic substitution and rationalisation of the directive effects exhibited by the methoxy and nitro substituents.]

167. Discuss the different mechanisms by which alkyl halides undergo reaction with typical nucleophiles.
[Your answer should include, but not necessarily be confined to, consideration of the variety of nucleophiles with which they react, and experimental investigations which have led to a complete description of the precise nature of the individual steps involved in these different mechanisms.]

168. Answer **ALL** parts.
(a) Draw an unambiguous perspectiwe structure of the *R* isomer of 1,1-dibromo-2-chloropropane, and briefly rationalise your answer.
(b) Suggest a synthesis of 4-Me_2N-C_6H_4-N=N-Ph from benzene.

(c) Write a structure for products **A** and **B**, obtained in the following reaction sequence and write a mechanism for the formation of each:

$$PhCO_2Et + CH_2(CO_2Et)_2 \xrightarrow[\text{EtOH}]{\text{NaOEt}} \mathbf{A} \quad (C_{14}H_{16}O_5)$$

$$\mathbf{A} \xrightarrow[\text{heat}]{\text{dil. HCl}} \mathbf{B} \quad (C_9H_8O_3)$$

169. A compound **A**, C_4H_8O, gives an orange precipitate on treatment with 2,4-dinitrophenylhydrazine. Treatment of **A** with ammoniacal silver nitrate produces a silver mirror, and converts **A** into compound **B**, $C_4H_8O_2$, which is soluble in aqueous NaOH. The compound **B** can also be obtained by successive treatment of propene with HBr, NaCN and aqueous acid. Compound **A** reacts with sodium borohydride to give **C**, $C_4H_{10}O$. On treatment with acid, **C** dehydrates to give an alkene, **D**, C_4H_8. What are the structures of **A** to **D**?

170. The reaction of bromine with an alkene may be used to show the presence of a double bond in a compound.

(a) Outline the mechanism by which bromination of a double bond occurs.

(b) How many grams of bromine will react with 7.0 g of pent-1-ene?

(c) What is the molecular weight of the alkene, 2.24 g of which rea ts with 3.20 g of bromine?

(d) When bromination of an alkene is carried out in water the product is a bromohydrin. Write a mechanism for bromohydrin formation.

(e) When a bromohydrin is treated with NaOH, an organic product, NaBr and H_2O are produced. Suggest a structure for the organic product.

REFERENCES

Aitken, R. A. and Kilenyi, S. N. (1994) *Asymmetric Synthesis*, Blackie Academic and Professional.

Cartmell, E. and Fowles, G. W. A. (1979) *Valency and Molecular Structure*, Butterworth, Fourth edition.

Corey, E. J. and Cheng, X.-M. (1989) *The Logic of Chemical Synthesis*, John Wiley and Sons.

Edenborough, M. (1994) *Writing Organic Reaction Mechanisms: A Practical Guide*, Taylor and Francis.

Field, L. D., Sternhell, S. and Kalman, J. R. (1995) *Organic Structures from Spectra*, Second edition, John Wiley and Sons.

Gray, H. B. (1973) *Chemical Bonds: An Introduction to Atomic and Molecular Structure*, W. A. Benjamin Inc.

Gunstone, F. D. (1974) *Programmes in Organic Chemistry, Volume* 8: *Basic Stereochemistry*, The English Universities Press Ltd.

March, J. (1992) *Advanced Organic Chemistry: Reactions, Mechanisms, and Structure*, Fourth edition, John Wiley and Sons.

Marples, B. A. (1981) *elementary Organic Stereochemistry and Conformational Analysis* (Monographs for Teachers Number 34), Royal Society of Chemistry.

Scudder, P. H. (1992) *Electron Flow in Organic Chemistry*, John Wiley and Sons.

Simpson, P. (1994) *Basic Concepts in Organic Chemistry: A Programmed Learning Approach*, Chapman and Hall.

Streitwieser, A., Heathcock, C. H. and Kosower, E. M. (1992) *Introduction to Organic Chemistry*, Fourth edition, Macmillan.

Sykes, P. (1984) *A Guidebook to Mechanism in Organic Chemistry*, Fifth edition, Longman.

Warren, S. (1974) *Chemistry of the Carbonyl Group: A Programmed Approach to Organic Reaction Mechanisms*, John Wiley and Sons.

Warren, S. (1978) *Designing Organic Synthesis: A Programmed Introduction to the Synthon Approach*, John Wiley and Sons.

Williams, D. H. and Fleming, I. (1995) *Spectroscopic Methods in Organic Chemistry*, Fifth edition, McGraw-Hill.

INDEX